1001 IDEAS FOR
FLOORS

FLOORING SOLUTIONS FOR EVERY ROOM

1001 IDEAS FOR
FLOORS

FLOORING SOLUTIONS FOR EVERY ROOM

EMMA CALLERY

First published in the UK in 2008 by
Apple Press Ltd

Apple Press Ltd
7 Greenland Street
London NW1 0ND
www.apple-press.com

A Marshall Edition
Conceived, edited, and designed by
Marshall Editions
The Old Brewery
6 Blundell Street
London N7 9BH
U.K.
www.quarto.com

ISBN: 978-1-84543-224-9

Current printing (last digit)
10 9 8 7 6 5 4 3 2 1

Originated in Hong Kong by Modern Age
Printed and bound in China

Publisher: Richard Green
Commissioning editor: Claudia Martin
Art director: Ivo Marloh
Design and editorial: Seagull Design
Illustrator: Mark Franklin
Project editor: Amy Head
Indexer: Lisa Footit
Production: Nikki Ingram

Front cover photo: Kährs
Front flap photo: Peter Woloszynski/
Redcover.com
Back cover photo: Meridia Meridian
Back flap photo: Richard Powers/
Redcover.com

Contents

Introduction

Floors are the hardest-working surfaces in your home. They are expected to cope with the pounding of innumerable footsteps, plus take the drops and spills of everyday life. Not only do they work hard, but we want them to look good, too. After all, floors can make the difference between a dull and a sparkling room. It's a lot to ask, really, isn't it?

The good news is that today's flooring materials are up to the task, and they are available in an incredible range for us to choose from. Who'd have thought that wood would make such a comeback, or that it would become fashionable and practical to use materials such as cork and concrete for flooring? While traditional fitted carpets still have an important role to play, the current focus is on hardwood, ceramic and stone, all of which offer both elegance and a broad variety of colours and textures. Another advantage is that, thanks to technological improvements, these flooring materials are easier than ever to buy and install.

The choice is so wide that you can go beyond simple practicality in your choices and find flooring to suit the house, the individual room and your lifestyle. Your first step should be to identify the materials that are compatible with your house. A house is part of an environment, and if it doesn't incorporate local materials it can seem out of place.

Also consider the style of the building. What are the design guidelines? For example, an American colonial-style home might boast stained or painted softwood flooring with stencilled borders, topped by rugs. An English country-style home would include stone or wood, while the Victorian era favoured fitted floral carpets and quarry tile or parquet flooring. An arts-and-crafts home would feature hardwood floors with intricate inlays, scattered with patterned rugs. The contemporary stylist might consider exotic woods, natural looks in stone, or ultra-modern materials such as rubber or terrazzo.

Such guidelines do not have to be strictly followed. You can choose which elements to include and evolve your own eclectic style that suits your character and lifestyle, as well as your environment and the existing or planned décor. However, adjacent rooms should be in harmony, so that the transition between different parts of the house is not jarring.

The next step is to consider the needs of each room, in terms of practicality and style: no one has a shag carpet in their kitchen! Do you want the room to have radiant heating? What colours, patterns and textures would suit the space? Are there changes of level to highlight or mask, or other areas to define? Who uses the room, and is it shared with any pets? Think it through carefully.

Opposite: Clean lines of glass and metal blend with warm, natural beech floorboards in this stunning contemporary setting.

Below: Nothing matches the comfort of carpet, and there are natural forms of this material that will please those who are environmentally aware.

①

Durability is one consideration. High-traffic rooms, such as entrance halls and kitchens, take a daily pounding, and some materials (lesser quality carpeting, for example) can't cope. Floors are tricky and expensive to install, and you don't want to have to repeat the laborious process every couple of years.

There are other factors that could particularly concern you. For safety, you might install nonslip bathrooms and kitchens and ensure a suitable environment for young children or people with limited mobility. For comfort (those who favour bare feet or lying down in front of the television), incorporate a carpet or rugs. Think about sound, too, particularly televisions, sound systems and the floor itself. Footsteps on a wooden floor in a large room can echo like they would in a crowded dance hall.

Budget is also a key element. There is an enormous variation between, say, a marble or leather floor and the least expensive vinyl or synthetic carpets. If your heart is set on stone or wood but your pockets are not deep enough, there are some great imitations around. Remember to allow for installation costs as well as materials.

Maybe you want to install the floor yourself. Resilient floors are relatively easy to put in, but they are the

②

③

① Rugs are great room softeners; here, making a rustic brick floor so much more welcoming.

② Meet the great impersonator: this stylish tiled floor is actually laminate (plastic), which can be laid in hours rather than days.

③ In traditional settings, few materials can rival the warmth and aged character of oak.

④ For a surface that will last a lifetime, go ultra-modern with concrete. Here it creates a bold, striking effect.

④

thinnest type of flooring. This means that the subfloor must be absolutely level, because any flaws quickly show through. Tiles are another flooring type that DIYers can manage, but the process – planning, cutting and installing – can be time-consuming. Floating wood floors don't necessarily require professional installation, but glued and nailed floors are another matter. Other flooring types – hard varieties such as granite, marble and terrazzo and soft flooring, such as carpet – require special installation techniques most familiar to those in the flooring trade. They are probably best left to the experts!

Currently, many interior designers are favouring environmentally friendly flooring. This rules out plastics, such as vinyl and laminates, but allows natural products to shine, such as stone, sustainably harvested woods (including cork and bamboo), natural soft flooring (such as sisal) and, perhaps surprisingly, linoleum, which is manufactured from a mixture of natural products.

Choosing flooring is a demanding process that shouldn't be rushed. Study your friends' floors, visit showrooms, bring back samples to view in natural light and use this book to help you discover your dream floors.

Matching Floors To Rooms

Consider each room and its requirements. Small rooms call for small patterns, such as parquet, while larger rooms have space for bigger shapes, like stone slabs. Dark floors define space well, while pale or neutral tones deflect attention elsewhere, making the room seem bigger. Carpets and matte finishes absorb light, confining the scale of a room, while shiny surfaces create a brighter, airier look. Floors are a key element in the overall decorative design of a room, as shown here.

②

③

①

① This polka-dot vinyl floor is very practical for a kitchen, and its brightness maximises the amount of light and enlivens the subdued tones of the room.

② Natural rugs are excellent at adding warmth and texture. This is especially valuable in contemporary settings.

③ Carpets and rugs offer colour and pattern, plus a cosy landing for bare feet.

④ Exotic hardwoods, such as this jatobá, or Brazilian cherry, are proving popular in bedrooms. The look here is chic and striking, but still welcoming.

⑤ If a room links with the exterior, make a link in the décor. This looks like stained wood, but actually it's linoleum – a floor made with natural ingredients.

Living Rooms

The priority here is comfort, but consider how the room is used. A busy family room might need a dense carpet made of nylon, which stands up to anything. If the design scheme is to be formal and luxurious, go for a smooth Saxony carpet or, for the highest level of comfort, a deep shag carpet. The options go much further than carpet, however. Wood can be wonderfully warm and welcoming (especially when combined with rugs), while stone creates an impressive setting. Modern linoleum, leather, or even concrete can add true style to your home.

① Linoleum has made a major comeback, even for living rooms, because it feels soft underfoot, it's warm and it offers many options.

② The coarse grain of oak brings warmth and style. Here it has been laid in strips to elongate the room. The rug adds texture and defines the sitting space.

③ This room's blend of rustic and modern calls for an unusual floor, which is provided by random-shaped stone tiles.

④ Do you want contemporary cool? Try concrete for a striking and surprisingly welcoming look.

⑤ Carpets are the usual choice for living rooms. It's best to balance bold patterns with neutral décor.

⑥ Rugs can be used with great flexibility to bring comfort and define space where needed.

Dining Rooms

Dining rooms are used for entertaining. They need to be stylish and intimate, but with a floor that is easy to clean and resistant to stains. If you opt for carpet, avoid a deep pile. Just about any other surface is an option, so the key is to select flooring that complements the rest of the décor and will allow your food and company to shine. Don't forget, you are creating a setting. On a practical note, remember that chairs can create a cacophony as they scrape across stone or wooden floors.

②

①

① A rug is great for defining dining space – as long as it is easy to clean. Here, a rug is paired with an elegant, polished stone floor.

② The stone-chip finish of terrazzo is grand and durable, but more expensive than many other options.

③ If you want a carpet that is both stylish and practical in a dining area, go for natural materials. They do not have a dense pile so they are easier to clean.

④ A hickory floor sets the stage for memorable dining in this dramatic room.

⑤ Once a kitchen staple, linoleum copes well with the stains and spills of dining.

⑥ A laminate wood-effect floor will make less noise than the real thing when these stools are being repositioned.

Kitchens

Aside from carpet, you have carte blanche in choosing from the flooring menu for a kitchen. Stain- and water-resistance, ease of cleaning and hygiene are priorities, and just as important is a smooth, nonslip surface. Bear in mind that dirt can collect in the gaps between wooden floorboards and sunken grout. Also consider the length of time people tend to spend on their feet in the kitchen. A soft and forgiving resilient floor may be more comfortable than hard stone slabs – it will also reduce the likelihood of any dropped objects shattering.

① Woods, such as this oak, will match other wooden finishes, but remember that they must be sealed against moisture.

② Stone tiles are a practical and stylish option and cope well with changes of level like these steps.

③ Often labelled 'cheap but practical', vinyl is a real winner in the kitchen because it adds colour and copes well with spills and stains.

④ Stone brings a room back to natural elements and is perfect for rustic kitchens.

⑤ Linoleum was the default kitchen flooring for years, and is now back in favour, partially because of its rich decorative possibilities.

Bathrooms

The Romans knew how to build bathrooms, and it is hard not to agree with them that ceramic tiles are the perfect bathroom floor: they are easy to clean, water- and slip-resistant and capable of creating a decorative element that will distract from the bulk of the bath or shower cubicle. For those who want to bathe in grand style, marble is the classic material. Yet there are many materials that challenge the precedence of both ceramics and marble. Here we explore some of the options.

① An interesting newcomer is the glass floor, which complements other bathroom fittings well.

② For a clean, subdued effect, this limestone floor is hard to beat, especially when paired with decorative pebbles.

③ Marble makes a grand statement wherever it is. This bathroom is shouting 'Look at me!'.

④ You can use wooden flooring material in bathrooms, but certain precautions should be taken to avoid rotting. Here, the boards have been set directly onto the joists, allowing the air to circulate and preventing moisture from damaging the planks.

⑤ Ceramic tile is the traditional bathroom choice because it looks good and performs well.

Hallways

Hallway floors should be welcoming as well as able to cope with heavy traffic. Because many hallways extend from the front entrance, they tend to attract dirt from outside. With this in mind, a surface that can be brushed clean is easier to live with than carpet. To add warmth to stone finishes consider incorporating a rug that can be easily cleaned. Scale is a factor, too: small hallways need pale colours in plain or intricate patterns. Above all, the hallway must complement the flooring designs in adjoining rooms. With all of these considerations, it's no wonder the classic Victorian tiled floor is hard to beat!

① Marble is a favourite choice in hotel lobbies because it is stylish and practical: a wonderful choice if your hall and budget can take it.

② Set stripes across narrow hallways to make the space seem wider. With carpets, go for a shallow, hard-wearing pile.

③ A strikingly modern glass hallway is shown here. Its clean lines have been softened by potted plants.

④ Wood can be both formal and welcoming, and here it provides a link with the natural world outside.

⑤ Linoleum flooring is sufficiently hard-wearing for use in hallways. Here, the bright sand finish reflects plenty of light, making a relatively narrow space seem very roomy.

⑥ Almost any other material would be overshadowed in this setting, but a geometric linoleum design leads visitors into the house in style.

Staircases

Staircases mark transitions between parts of the house and should echo the décor at one end (perhaps in material or colour) so that the passage is smooth. Neutral designs usually work best. Stairs are easily worn down by heavy use, so you need a highly durable material. The most important concerns are safety and noise: the surface must be nonslip and must soften the footfalls of its users – a carpet or a runner will do the job well. Hard tiles and vinyl should definitely be avoided.

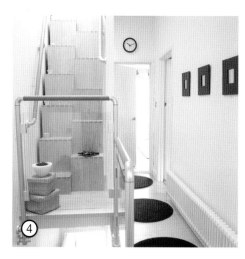

① Keep stair carpets neutral to avoid clashes between different colours or surfaces.

② A concrete spiral staircase turns a functional need into an architectural design feature.

③ Stone staircases work well for entrance halls that link directly to the material used outside.

④ This wooden box staircase has a sculptural quality, making it a great focal point.

⑤ By pairing wood and metal in this staircase, a decorative link is established between the wooden floor and the room's contemporary styling.

Bedrooms

The bedroom is a private space where you have the chance to indulge yourself in terms of comfort or style – ideally, both! Bedrooms get less traffic, so you are free to use a special colour or texture that doesn't have to match any other part of the house. Comfort and sound insulation are the main requirements, which is partially why fitted carpets have proved to be enduringly popular. Avoid the prickliness of natural fibres and the artificial feel of laminates.

① In an exotic setting these tiles make an enormous contribution to the romantic, Eastern atmosphere.

② Stone isn't the warmest material, but for a cool bedroom with many points of interest, it really fits the bill.

③ Jatobá laminate matches the wooden bed and contemporary style of this bedroom, which succeeds in being stylish but welcoming.

④ The indulgent, luxurious furnishings of this bedroom are matched by the mocha-stained, red cherry floor.

⑤ Plan the room as a whole: this ornate bed and drape coordinates with matching colours in the carpet.

Children's Rooms

When choosing flooring for a child's bedroom, balance practicality with personality. You want an easy-to-maintain floor that is forgiving on toys and knees so that it is a welcoming play area and can be easily cleaned (if you are lucky enough to catch sight of the floor itself!). Bright colours are tempting, but they often become dated very quickly. If you opt for carpet, go for dense, level loop textures and high wool content to minimise those painful carpet burns. Wood is a practical choice that can be softened by rugs.

②

①

① Oak flooring makes a fine surface for toy vehicles and is very easy to sweep clean.

② Linoleum offers great practical advantages combined with a vast range of colours and designs.

③ If budget is an issue, laminate offers all the joys and none of the pains of wood.

④ For a playroom, ceramic tiles come in numerous colours and are very easy to clean.

⑤ Many princes and princesses crave comfort in their bedroom, in which case carpet is the best option.

Stone Floors

Stone is an excellent flooring material because it is beautiful and durable. It can take the pounding of decades of footsteps, and its looks improve with time. Stone is also versatile. There is such an incredible assortment of colours, patterns and textures that, if your budget allows, it will complement almost any setting. While granite is the most popular option, the range is enormous and varies widely in cost.

Stone is one of the oldest materials on our planet. First formed with the Earth's crust four billion years ago, it has been used for buildings and tools by countless civilizations. In addition to its beauty, stone has a sense of permanence and offers connections to history and our natural environment.

Stone's tough characteristics make it ideal for use in heavily used areas, such as entrances and hallways, but it is also suitable for kitchens and bathrooms – and it can create a stunning living room. Once only available in heavy slabs, stone can now be found more easily in thin tiles, which are easier to handle. Despite this, installing stone remains a job for professionals.

This section's illustrations show the incredible variety of finishes available, giving each stone type a distinct style. Some stone bursts with veins and streaks, even embedded fossils, while other kinds offer a cool, calm look. Colours vary significantly, from the palest cream to the deepest black, taking in a palette of greys, yellows, blues and greens along the way.

One drawback of a stone floor is that it feels cold – great for cooling a room during summer and for natural refrigeration in a pantry, but elsewhere it can benefit from underfloor (radiant) heating to warm the surface. Stone's hardness also makes it tough on feet and liable to damage anything that has the misfortune to fall on it. It is important to beware of this because stone can become slippery when it is wet. Also remember that stone must be sealed to maintain its durability, especially if it is subject to moisture and spills, as in a kitchen or bathroom. These are all minor, manageable concerns, considering that stone adds beauty and elegance wherever it is used. And at the same time you can be certain that your stone floors will stand the greatest challenge of all: the test of time.

Opposite: Rustic is the key look in this living area where aged stone, wood and leather are the main materials. The soft furnishings add colour accents.

Below: Slate tiles bring a sense of elegant drama to a room.

Types and Finishes

The main types of stone flooring are marble, granite, limestone and sandstone. They are sold as slabs in various sizes or as tiles, which are often smaller, thinner and lighter. All stone requires a strong subfloor that can bear considerable weight. Freshly quarried stone does not have the charm of age, but the finish can be roughened into a more worn effect by a process called 'tumbling', in which the surface is blasted with small particles of limestone. Smooth surfaces can be honed into a flat matte or low-sheen gloss. All of these processes make the stone more porous and, therefore, more susceptible to staining. Proper sealing and regular maintenance is all that's necessary to keep a stone floor in tip-top shape.

① Marble comes in many colours, perhaps the warmest of which include these shades of pink.

② The naturally shiny finish of slate balances the dark colour of this floor and adds a touch of the outdoors to this contemporary setting.

③ For a cool, minimalist look, choose limestone.

④ Flagstones are a reminder that stone is the most basic of materials, forming the natural ground beneath our feet.

⑤ For a floor with unique decorative interest, choose from the many styles of mosaic that are available, or design your own.

④

⑤

Sizes and Proportion

When choosing your material, bear in mind the sizes of the tiles or slabs supplied; size has a major impact on the final effect. Larger slabs, especially if they are irregular, give a rustic look – but they can have associations with outdoor pavement, which may seem odd within a domestic setting. Grout width is another important consideration. Laying tiles edge to edge gives a smoother finish, but the patterns created by the lines of grout can be pleasing and can highlight the tile pattern.

① Use a mixture of tile sizes for a warmer, welcoming look.

② The larger the room, the larger the size of tile it can take.

③ Smaller tiles give a busier feel – this will look better with a plain or smooth finish, where there is no pattern to compete with the grout lines.

①

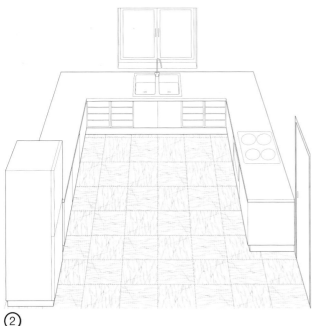

②

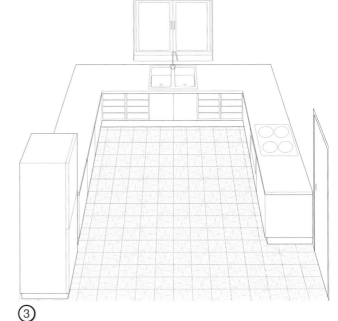

③

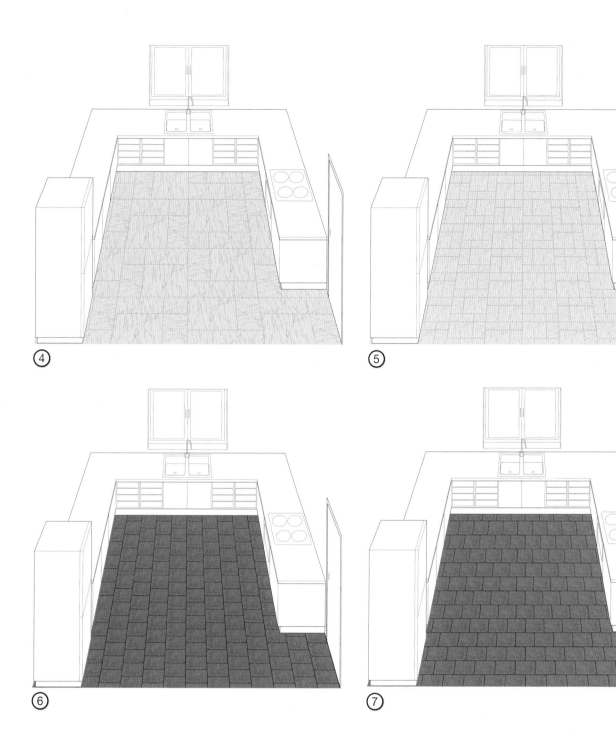

④ Avoid a uniform look by combining square and rectangular tiles that blend in with the room.

⑤ Small square and rectangular tiles create an attractive pattern and make a small space appear larger.

⑥ Consider the direction in which you lay your tile. Straight lines lead the eye, so a room with lines down its length appears elongated.

⑦ Lines going across the room will make the space seem shorter, but wider: a great option for long, narrow spaces.

Granite

Granite is an igneous rock formed by the slow cooling of magma (below ground) or lava (above) pockets. Its density differentiates it as the hardest and most resilient stone – impervious to water and highly resistant to wear and chemicals – so it is good for high-traffic areas and requires minimal maintenance, especially compared with some of its rivals. Day-to-day wear and tear adds to its rugged charm, but if it does get damaged, most scratches can be buffed out. For these reasons, granite is the most popular choice for stone flooring. However, remember that granite is heavy, and your floor must be able to support its considerable weight.

① Granite can be almost pink, as seen here. In this grand bathroom the matte-finish floor contrasts with the beautiful sheen of the bathtub and surround.

② Mosaic insets separate these granite tiles, creating a striking and stimulating finish.

③ The flecked finish in this granite-lined bathroom creates a sense of energy and vivacity.

④ Interest and character have been added to the floor at the base of this staircase by using several shades of polished granite and incorporating an inlay.

③

④

Granite comes in a spectrum of colours, from near-black to mottled white, incorporating browns, yellows, pinks and blues. Beyond its base colour, granite contains flecks of mica, quartz and feldspar – the minerals that make up much of the Earth's crust. The speckled patterns created by these imperfections give granite its majestic, coarse-grained character. Go for rough textures, as good grip makes them most suitable for flooring. This texture can also be achieved artificially by a process called 'flame texturing'.

① Mottled brown
② Grey and blue
③ Deep brown and grey
④ Golden yellow
⑤ Speckled gold
⑥ Storm-cloud grey
⑦ Polished gold
⑧ Black and grey
⑨ Deep cream
⑩ Deep blue and brown
⑪ Sea green
⑫ Deep blue

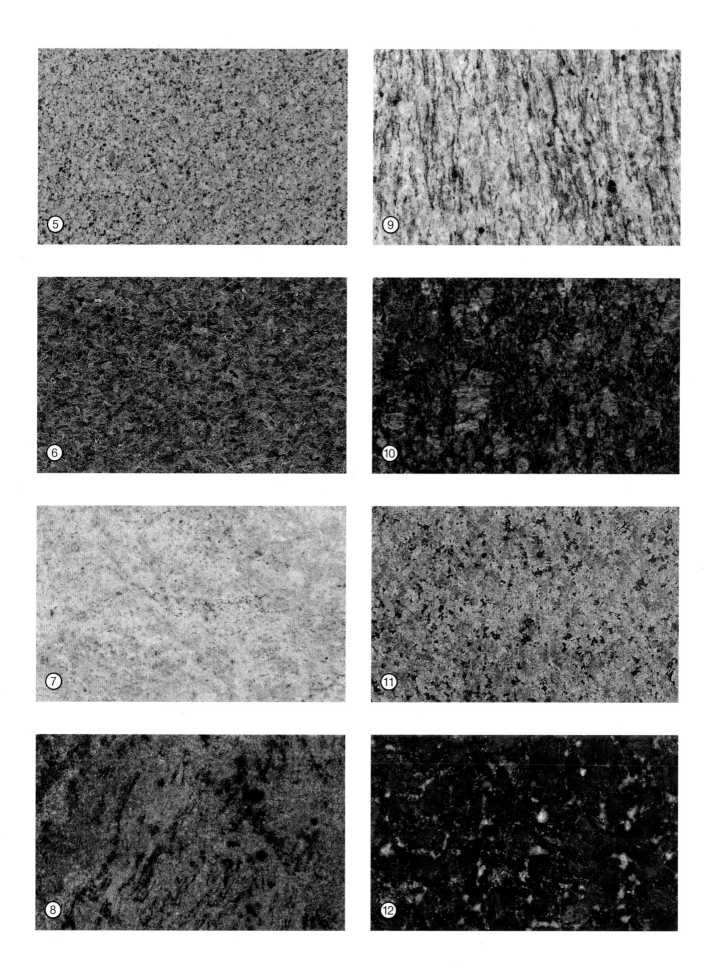

Limestone

Limestone is born from water, as it is mostly formed from a mineral called calcite, which is made up of the crushed remnants of marine organisms and sediment. It is commonly used to complement chic, minimalist decorative schemes. However, the range of colours available – from creamy white to dappled blue – makes limestone suitable for many styles. Its rustic texture adds to this versatility: the wooden furniture of a country-style kitchen is complemented beautifully by the timeless appeal of aged flagstones. Limestone is softer than other rocks, such as granite, so it must be sealed to keep out moisture, and it requires regular maintenance. Some people find that this softness makes it more comfortable underfoot than other stone materials.

(4)

① The cool elegance of this honed limestone complements its minimalist setting to great effect.

② This kitchen floor is laid in Cotswold stone, a honey-hued limestone that glows with the rustic character of its origins in the Midlands.

③ Laying a bathroom floor in limestone requires extra care in sealing and maintaining the material, which can be easily stained by water.

④ Link two areas of the home by using the same flooring for both – here, grey limestone unites the cooking and eating areas.

⑤ Black diamond insets break up the smooth uniformity of this honed limestone floor, adding visual interest to its cool style.

⑥ Limestone has many moods: its golden hue echoes the warmth of the exposed stone walls in this living room.

(5)

(6)

Limestone typically has a pale tone, given by the presence of minerals such as dolomite and aragonite, but impurities such as sand, clay and iron oxide create attractive grains, veins, speckles, or flecks. This patterning brings a sense of richness and permanence. The finish can be polished (for harder varieties honed to a matte surface or a slight sheen), brushed, or water-worn for a softer look.

① Antique-brushed dark grey
② Flame-finished grey
③ Brushed, black and blue
④ Antique-brushed cream
⑤ Tumbled cream
⑥ Black and blue swirls
⑦ Brushed amber
⑧ Honed grey with flecks
⑨ Polished with gold veins
⑩ Brushed, blue and green
⑪ Honed, tan and gold
⑫ Polished Egyptian gold
⑬ Cool grey
⑭ Honed sea grey
⑮ Honed rich cream
⑯ Tumbled 'cloudburst'
⑰ Honed 'mottled cloud'
⑱ Tumbled 'sunray'
⑲ Polished gold
⑳ Tumbled 'soft sky'

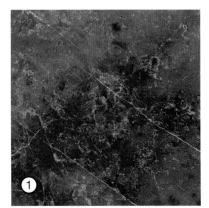

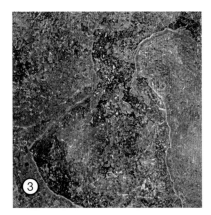

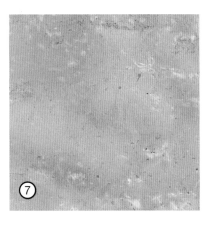

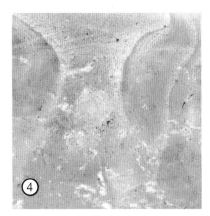

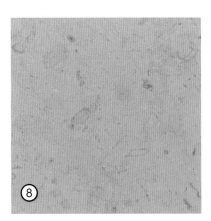

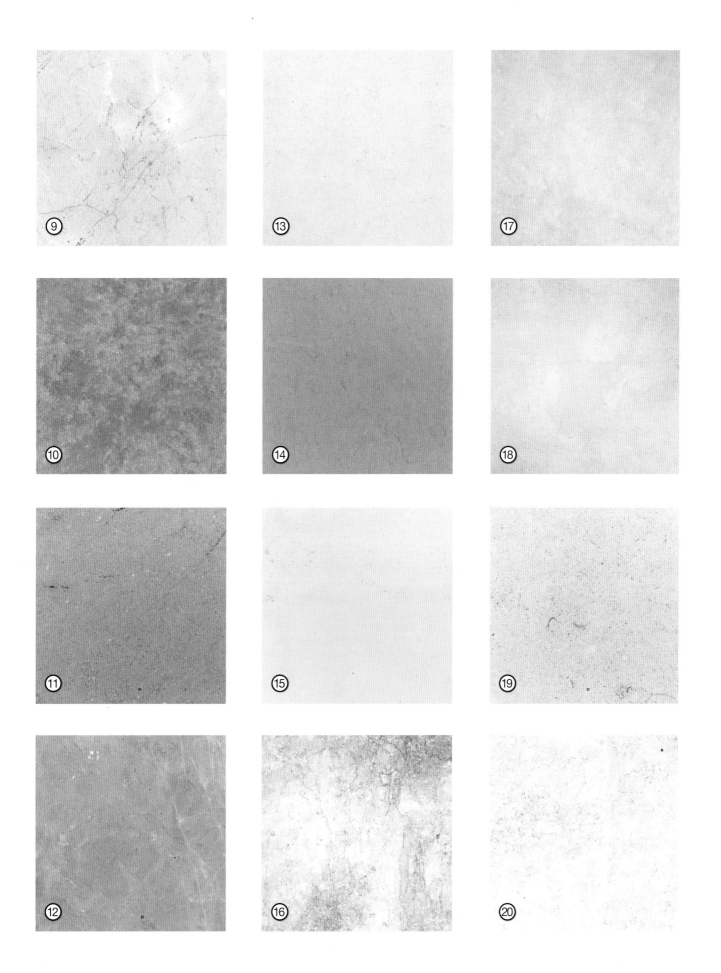

Marble

In the right setting marble can be spectacular. It carries with its considerable weight a sense of grand elegance that is popular in expansive settings, such as the foyer of an opulent hotel. To fit into a domestic decorative scheme it needs exactly the right setting and sense of proportion. Anything but understated, it calls for large, airy rooms or a formal setting, such as a classic, elegantly furnished bathroom. Marble is a cool stone, with subtle patterns showing through its almost translucent surface. It is not waterproof and is easily scratched or stained, so it requires careful maintenance. Honed marble with a matte finish is the most suitable for flooring because it is less slippery than a shiny, polished surface. The best marble is quarried in Italy.

① Marble suits a grand setting. Here it is at home with a high ceiling, pillars and a grand piano.

② Marble is popular for bathrooms that make a statement. Its expense may limit its use to smaller rooms.

③ A marble bathroom floor with a sunken shower creates a contemporary room with a luxurious twist.

④ You can make marble seem more intimate by using it in tile form (here, in a tumbled, antique finish), where its grandeur is tempered by the smaller scale.

⑤ Marble provides a counterbalance to strong architectural features, in this case ensuring that the fire surround does not dominate.

Marble is actually crystallised limestone and it is formed by the chemical changes that occur when that stone is subjected to sustained heat or pressure, such as the movement of the Earth's crust. It is available in many colours – from white (its purest form, favoured by classical sculptors) to brown – including vivid shades of green, blue and pink. It's also available in subtle patterns with random veins and mottling. Such variations are caused by the presence of impurities.

① Polished antique gold
② Polished cloudy grey
③ Polished Italian beige
④ Polished veined white
⑤ Polished pale mocha
⑥ Polished light tan
⑦ Polished mottled white
⑧ Polished cedar
⑨ Honed champagne gold
⑩ Polished classic grey
⑪ Honed gold
⑫ Polished jade
⑬ Polished cherry blossom
⑭ Polished beige, ivory and pink
⑮ Polished deep brown
⑯ Polished Inca gold
⑰ Polished black
⑱ Polished beige and ivory
⑲ Polished dark green
⑳ Polished terracotta

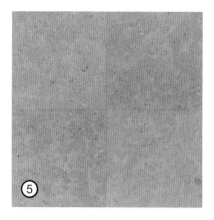

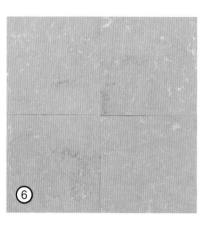

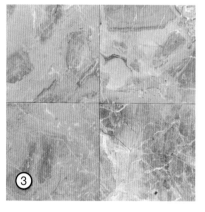

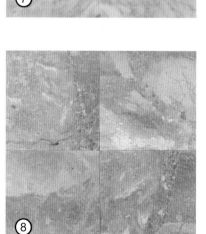

9

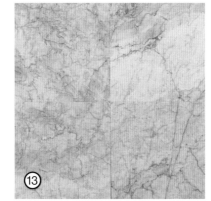

13

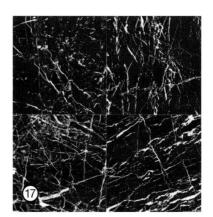

17

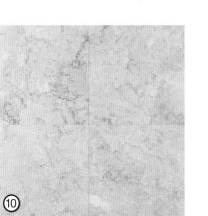

10

14

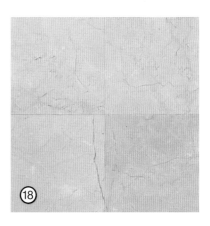

18

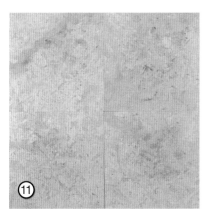

11

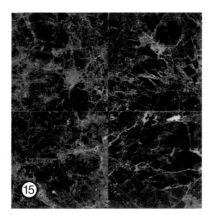

15

19

12

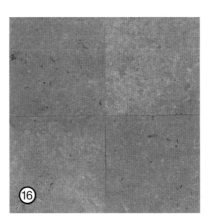

16

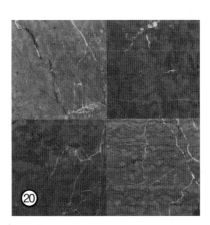

20

Travertine

What do the Colosseum in Rome and the Getty Center in Los Angeles have in common? The aesthetic finish of both relies on travertine, limestone's Italian cousin. Also known as travertine marble, this material is formed in the hot springs of Tuscany. Water percolates through limestone, releasing carbon dioxide gas like a carbonated drink and creating numerous cavities in the rock. The mottled, veined result is a popular and beautiful building and flooring material. Because it is pitted with holes it is porous and requires careful preparation and cleaning. A clear epoxy resin may be applied to create a smooth surface that is easier to maintain.

②

①

③

(4)

(5)

① Set travertine edge to edge with hardly any grout showing and it can look almost like marble.

② Irregular tiles of beige antique travertine reflect plenty of light – great for lifting a room where most of the furniture is dark.

③ Colour variation between stone tiles is not a bad thing, as seen here, with the extra texture offered by this bathroom floor.

④ Offsetting these rectangular tiles has created an interesting pattern that does not distract from the rest of this large entrance hall.

⑤ Travertine shows off its cool side in stylish, minimal settings like this one.

⑥ Travertine works very well with country-style colours and patterns, like those on the large pitcher and floral drapes.

(6)

Travertine has gas-created holes, which are usually filled in with resin or other fillers during the fabrication process, but the rock can also be tumbled to give it a slightly rough, old-world appearance. Although it is halfway between limestone and marble, travertine does not share these rocks' colour variety, offering a relatively narrow spectrum of yellows, greys and browns. There is no shortage of character, however, because of the numerous flecks, speckles and mottling that bring life to the surface of this intriguing stone.

① Filled light tan and grey
② Honed clouded cream
③ Honed and filled gold
④ Mediterranean ivory
⑤ Filled whirled cream
⑥ Unfilled coral
⑦ Polished and filled, vein-cut gold
⑧ Unfilled ivory
⑨ Honed mustard
⑩ Unfilled salt and pepper
⑪ Polished and filled, vein-cut cinnamon
⑫ Honed, filled light walnut
⑬ Filled dappled honey
⑭ Honed, filled cinnamon
⑮ Honed, unfilled light walnut
⑯ Antique-brushed, unfilled sea grey
⑰ Filled honey with streaks
⑱ Honed, unfilled cinnamon
⑲ Unfilled dark walnut
⑳ Polished and filled, vein-cut light walnut

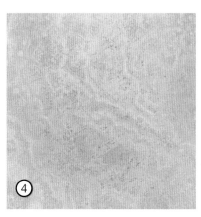

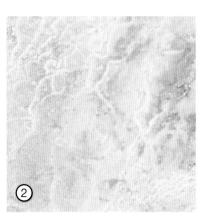

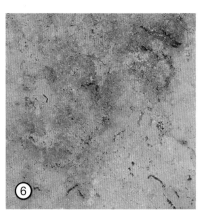

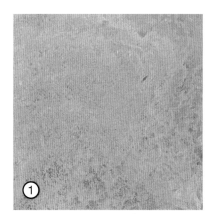

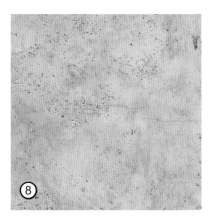

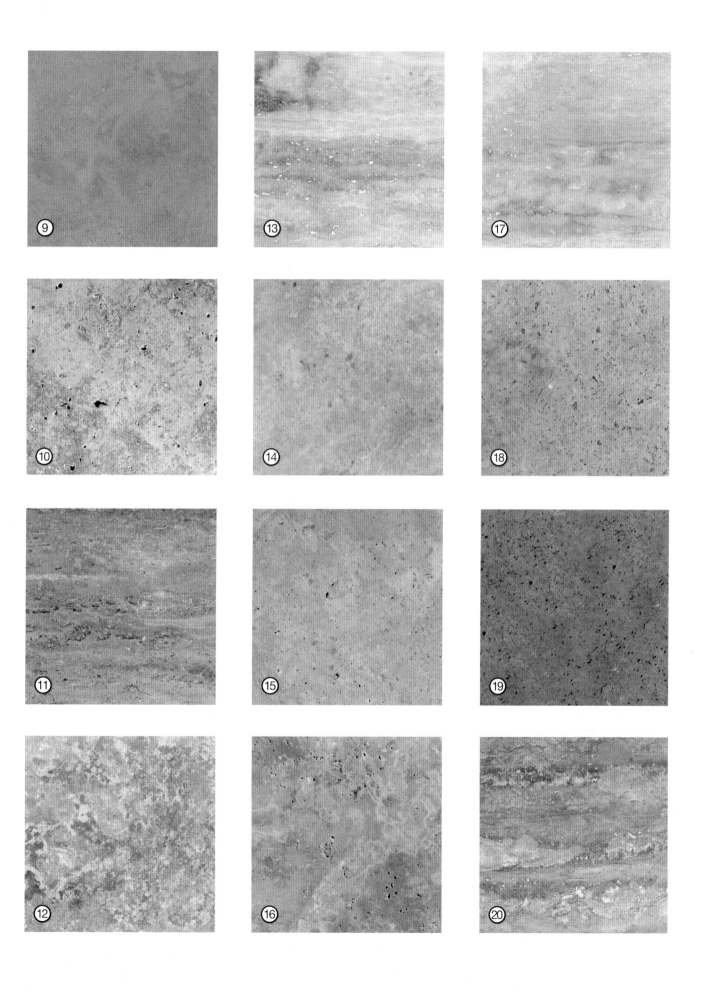

Sandstone

As its name suggests, sandstone is formed from countless compressed grains of sand, minerals and earth. Sandstone is attracting a lot of interest as a flooring material because it looks good, comes in many colours and is hardier, easier to maintain and cheaper than limestone, its fellow sedimentary rock. The structure and colour of the stone is determined by the proportion of minerals, such as quartz, mica, feldspar, clay and iron oxide. The hardest-wearing variety is York stone.

① Irregular setting of these rectangular slabs avoids overuniformity in this elegant grey sandstone floor which links the living room with the dining area.

② Some of the many hues of sandstone are cream and pink, here bringing a sense of warmth and solidity to this kitchen.

③ A smooth finish, a regular pattern and a subdued colour match sandstone to this simple, modern dining room.

④ Sandstone absolutely glows in natural light, as in this setting where a wealth of windows let in the sunlight.

Sandstone is available in the sandy colours of tan, brown, yellow, red, grey and white. The colour varies according to the source. For example, the sand of western USA is predominantly red, while Indian and Chinese sandstone is available in a massive range of hues. The rainbowlike array of vivid and subdued colours on these pages illustrates the enormous choice that is available in sandstone. Its variety of textures can also be very important as a design element.

① Natural white
② Natural grey
③ Natural with fossil
④ Natural red
⑤ Natural brown
⑥ Natural coral
⑦ Polished wheat
⑧ Natural pink
⑨ Natural pearl
⑩ Natural beige
⑪ Natural desert
⑫ Polished teak
⑬ Natural camel yellow
⑭ Natural yellow
⑮ Natural earth
⑯ Polished mauve
⑰ Natural chocolate
⑱ Natural mint green
⑲ Natural lilac
⑳ Polished gold with grey strips

① Natural white

⑤ Natural brown

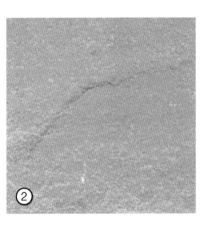

② Natural grey

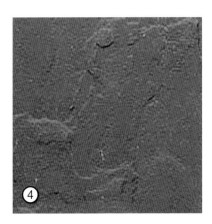

⑥ Natural coral

③ Natural with fossil

⑦ Polished wheat

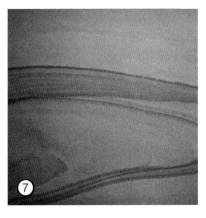

④ Natural red

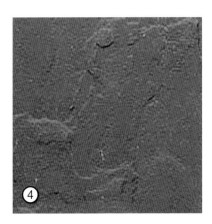

⑧ Natural pink

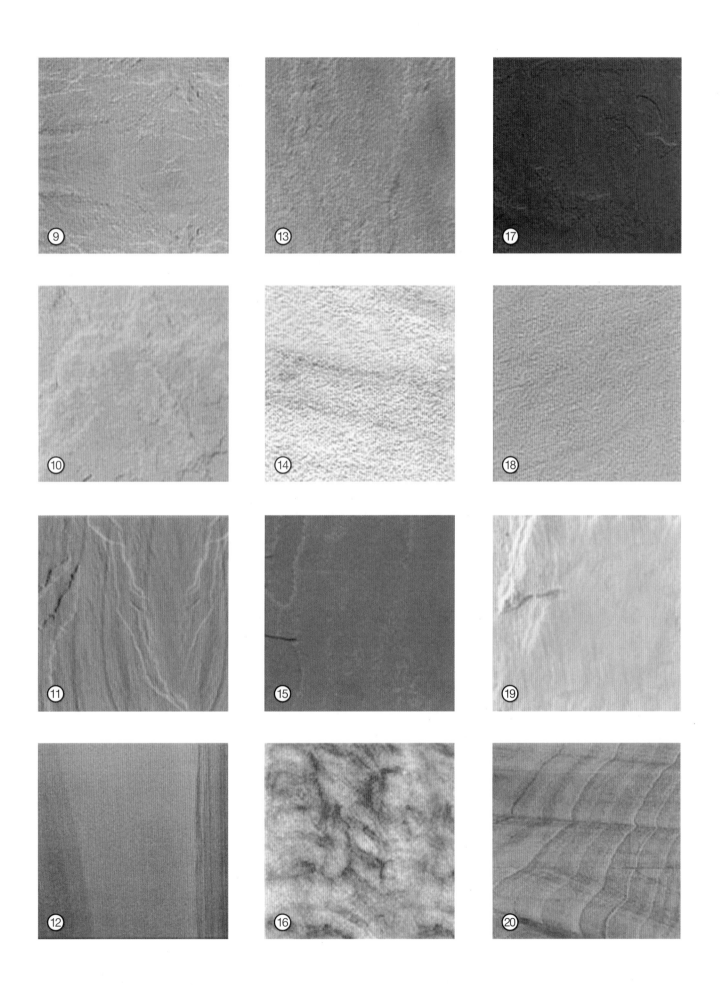

Slate

If you want your home to have a sense of spectacle and occasion, slate has as much drama as a theatre stage. Slate offers an amazing array of colours and patterns that can bring a room to life. This compact, hard and dense rock is formed under incredible pressure in the furnace of the Earth's crust, and it is now quarried in mountainous regions all over the globe. It won't flake or fade, it copes well with water and stains and it conceals dirt very well, making it ideal for kitchens, bathrooms and outdoor applications.

④

① Slate's intriguing texture is a definite advantage, adding interest to this grey-blue floor.

② Make a feature of slate's many hues by mixing them randomly, as in this busy conservatory floor.

③ The mixture of rust and grey-blue in this floor contrasts with the neutral backdrop of the kitchen.

④ Slate tiles in various hues of grey give this living area clarity, definition and understated class.

⑤ Jet-black slate set in a random pattern with white grout creates a showstopping floor, balancing the intense character of the ceiling.

⑤

Slate is popularly identified with just blacks and greys, but has so much more to offer: a true rainbow of oranges, reds and golds, as well as the metallic greens and blues. These colours often share space, blending into each other as they form what looks like an astronaut's view of Earth – this is fitting, since slate is so much a part of the planet's crust. The presence of mica minerals gives slate a naturally slick sheen.

① Riven silver-grey
② Riven barley
③ Honed Brazilian green
④ Honed blue and fawn
⑤ Honed Amazon green
⑥ Honed blue dawn
⑦ Honed sunset and grey
⑧ Riven cherry blossom
⑨ Honed dark grey
⑩ Honed sapphire contours
⑪ Riven lilac
⑫ Honed multicoloured
⑬ Riven copper
⑭ Honed pearl and rust
⑮ Riven silver and brown
⑯ Honed rustic grey
⑰ Honed grey
⑱ Honed ivory and gold
⑲ Sanded oyster
⑳ Riven earth

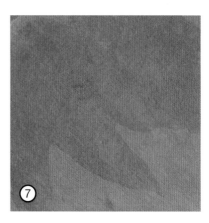

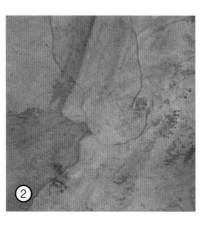

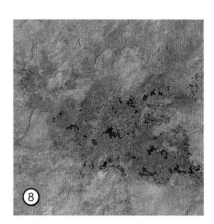

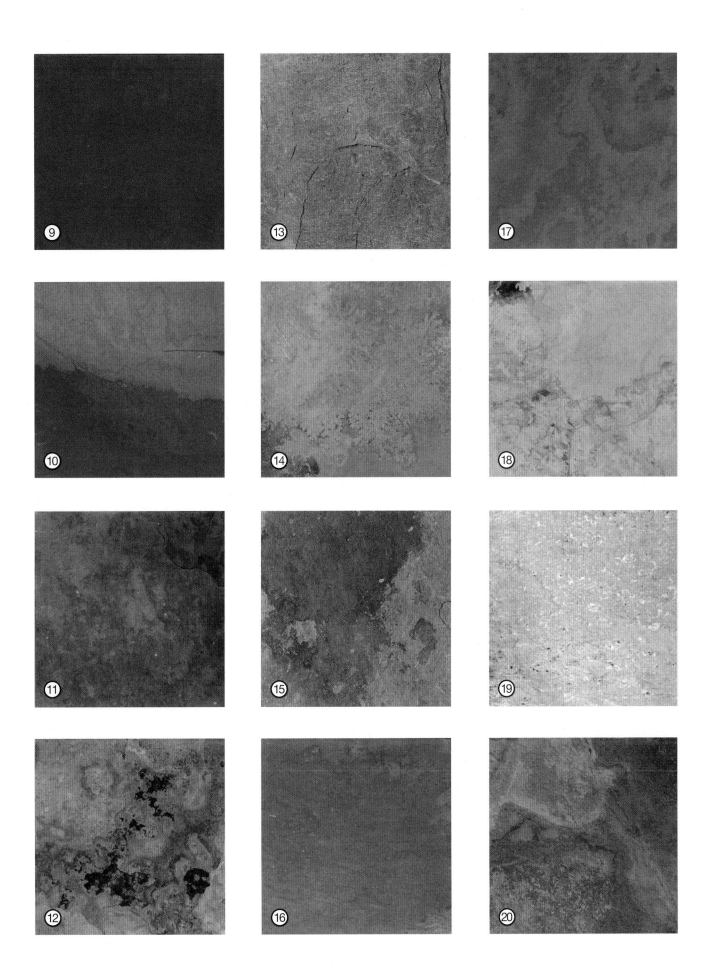

Mosaic

A mosaic floor takes you into the realms of the
Roman Empire. Romans were fond of this type of
covering because of its artistic potential. Mosaic is
made with small cubes that are bedded in mortar.
The cubes (or tesserae) can be formed in marble,
stone, glass, terracotta or coloured ceramic.
They are often available in sets of tiles attached
to a mesh backing that can be cut, if necessary,
and glued to the floor. The many lines of grout
are essential elements in the design and have a
practical value, since they reduce the slipperiness
of the surface, making it suitable for bathrooms,
especially shower (where the tiles can be easily
angled toward the drain).

③

④

① Small pieces of broken tile can be extremely effective in creating random patterns.

② Mosaic can be plain, as with these stone cubes. They resemble the wall tiles but offer variations in texture.

③ A mosaic pattern with an aquatic theme adds to the visual interest of this shower and increases slip-resistance.

④ This combination of white grout and neutral stone is calm yet interesting.

Mosaic designs can be plain, geometric or pictorial, and the method allows you to blend as many (or as few) colours as you wish into the decorative scheme. The size of the tile is a crucial decision: too large, and you lose the intimate effect of the sheer volume of tiles; too small, and the design looks busy and loses definition. One of the joys of mosaic is its flexibility: it can be used as a border or inlay, or to direct attention towards one part of the room.

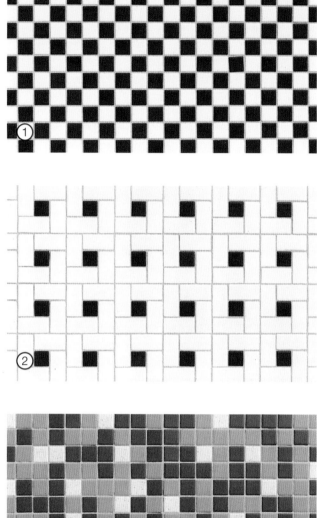

① Grouted tiles in classic checkerboard
② Grouted tiles in white around black
③ Grouted tiles in shades of blue and white
④ Grouted tiles in shades of brown
⑤ Grouted tiles in white textures
⑥ Grey mortar two-tone tiles, with wide joint
⑦ Cream, grey and gold tiles with wide joint
⑧ Rust and cream tiles, with wide joint
⑨ Grouted tiles in tan textures
⑩ Grouted tiles in sunlit grey
⑪ Grouted grey hexagonals and diamonds
⑫ Grouted tan hexagonals and diamonds

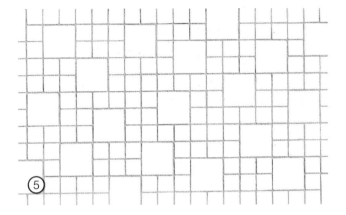

(5)

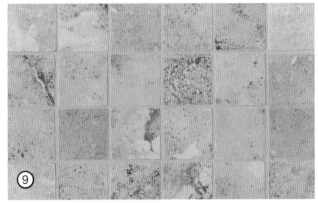

(9)

(6)

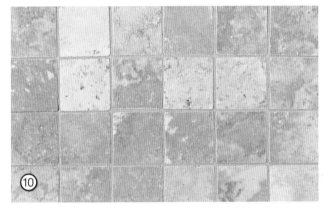

(10)

(7)

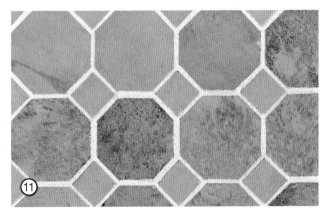

(11)

(8)

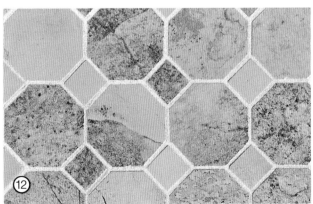

(12)

Continued from page 61.

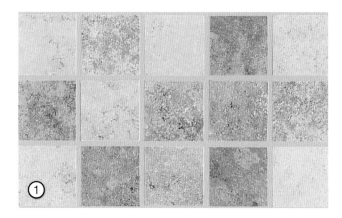

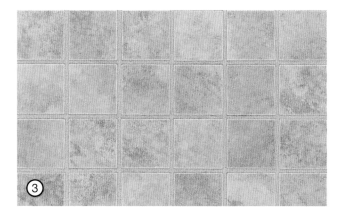

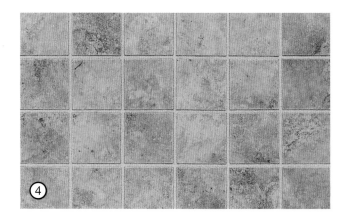

① Grouted tiles in grey shades
② Grouted tiles in fawn
③ Grouted tiles in tawny shades
④ Grouted tiles in chestnut
⑤ Grouted tiles in mist
⑥ Grouted contrasting hexagonals and squares
⑦ Grouted tiles in pale grey shades
⑧ Grouted tiles in ochre
⑨ Grouted tiles in golden yellow
⑩ Grouted tiles in ocean grey
⑪ Grouted tiles in the colours of spices
⑫ Loose sheet in a nut colour

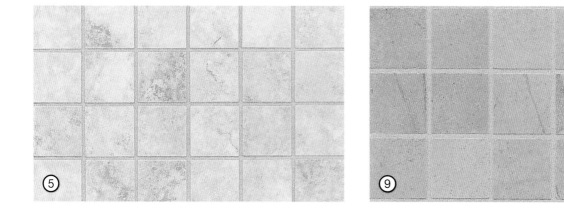

⑤

⑥

⑦

⑧

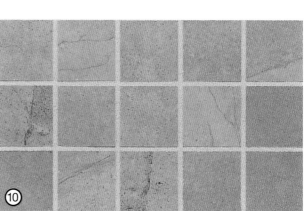

⑨

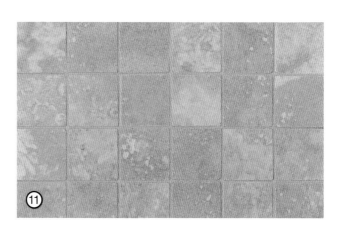

⑩

⑪

⑫

Continued from page 63.

① Loose sheet in golden shadow
② Loose sheet in russet
③ Loose sheet in light tan
④ Loose sheet in rust
⑤ Loose sheet in off-white
⑥ Loose sheet in forest shades
⑦ Loose sheet in clay
⑧ Loose sheet in white
⑨ Loose sheet in lagoon blue
⑩ Grouted tiles in chalk
⑪ Grouted tiles in ivory
⑫ Grouted tiles in lemon

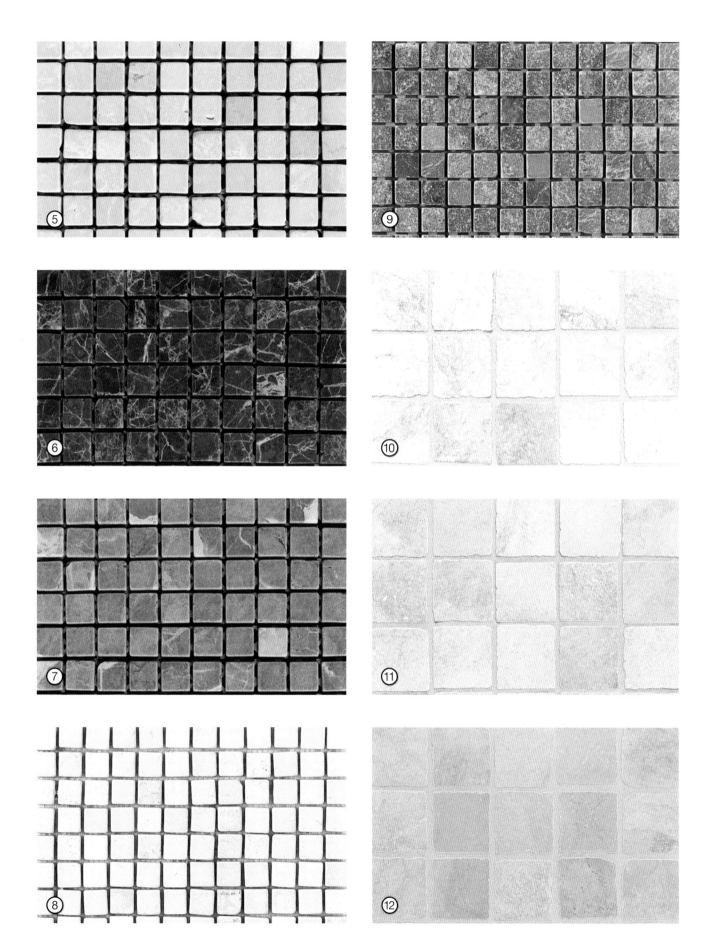

Continued from page 65.

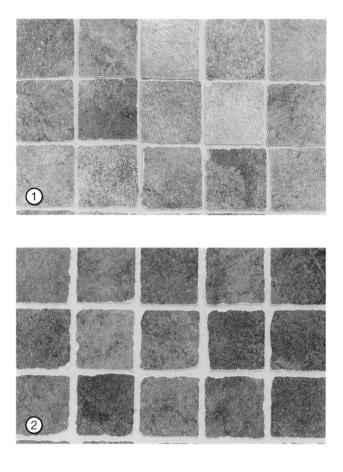

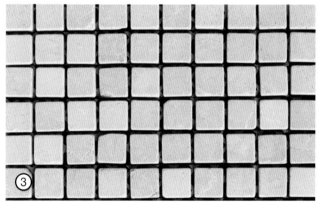

① Grouted tiles in dappled brown
② Grouted tiles in aqua grey
③ Loose sheet in rusted white
④ Loose sheet in off-white
⑤ Loose sheet in apricot
⑥ Loose sheet in stone
⑦ Loose sheet in powder blue
⑧ Loose sheet in pale grey
⑨ Loose sheet in straw
⑩ Loose sheet in peach
⑪ Grouted tiles in coffee and cream
⑫ Grouted tiles in blush

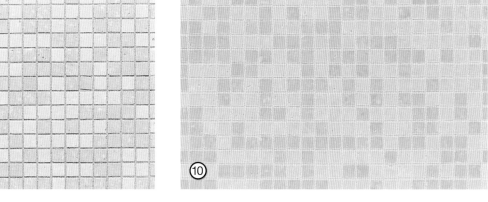

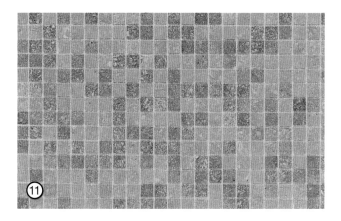

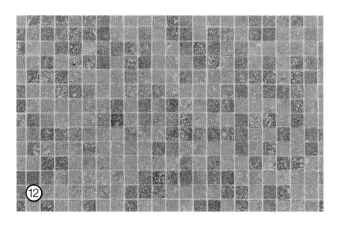

Terrazzo

Terrazzo is a combination of stone chips (usually marble or granite, sometimes with quartz or onyx) set in concrete, cement or resin. This method creates an extremely hard and stylish-looking surface that has proved to be very popular in commercial interiors such as shops and hotel lobbies. Also known as Venetian mosaic, it has been popular in Mediterranean countries for more than 100 years, partly because it helps keep houses cool in the hot climate. Terrazzo is now finding popularity in modern domestic settings, such as warehouse conversions and large entrance halls. In particular, it complements contemporary design elements such as chrome, glass and pale wood. However, this grand, luxurious material is almost as expensive as stone and must be installed by an experienced professional.

① Spectacular effects are possible with
terrazzo, as shown by this abstract design
in bold colours.

② Terrazzo can mimic other stone finishes.
This elegant blue-grey kitchen floor could
easily be slate.

③ Different levels in rooms like this one can
present challenges. The answer was a
terrazzo floor, which can handle the steps
and keep the look unified.

④ The stone-chip finish of terrazzo adds an
extra design element to this sophisticated
living-and-dining area.

⑤ A sleek effect – perfect for bathrooms – is
achieved here by combining white paint
and chrome with cool blue in the terrazzo
flooring and tile bath-surround.

Terrazzo has traditionally been mixed and laid on site, but it is now available as tiles or slabs. Once in place, it is virtually indestructible and, despite its smooth surface, it is not slippery at all. Almost any colour is possible, and in any combination, and it can be set to resemble a jumbled mosaic, geometric design or set in a motif of your choice. A cost-effective application is to use it for an inset strip to contrast with other materials.

① Designer mix: grey, gold and black
② Designer mix: earth colours
③ Exotic green, gold and black
④ Exotic blue flecks
⑤ Designer mix: pebbles
⑥ Designer mix: night sky
⑦ Exotic royal scarlet
⑧ Exotic shades of blue and white
⑨ Exotic purple and grey
⑩ Standard mist
⑪ Standard red, brown and gold
⑫ Standard charcoal
⑬ Copper and gold, rustic finish
⑭ Standard greys
⑮ Standard black and brown
⑯ Grey cobbles, Venetian finish
⑰ Standard white and grey
⑱ Standard fawn shades
⑲ Standard brass
⑳ Pea green and grey, Venetian finish

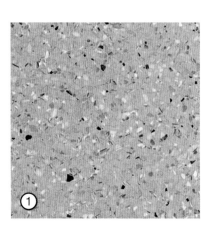

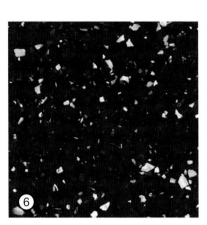

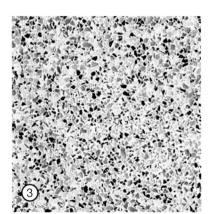

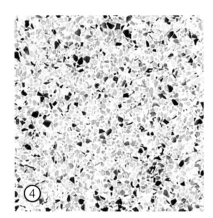

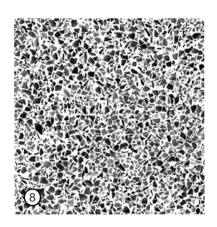

Using Colour

Dark colours bring the surface towards you – so darker walls make a room appear smaller, and lighter shades make it seem more airy. However, dark colours are warmer and more intimate than light shades. When choosing colours for a stone floor, consider the effect on the overall scheme in terms of colour, pattern, and texture. Stone tends to look and feel very cool, so in a large room, a dark floor may be more suitable. Pattern and texture draw attention, so a plain, honed floor is better in a small space where too much visual interest will make the room seem busy.

① Granite accompanies the exotic here when paired with citrus-hued lime-green walls. Since all the tones are colours from nature, the effect is still warm.

② The warm tones of these café au lait walls and deep brown granite create a cosy, welcoming environment.

③ Cream walls are easy on the eye and soften the hue of the floor by emphasising its lighter elements.

④ Pair brown granite with orange walls for a Mediterranean look. These two shades complement each other because they are near neighbours on the colour wheel.

⑤ This light shade of granite combines with cream décor for a soft, contemporary look.

⑥ Pastel shades such as this pale jade are fine companions for the wood and sand tones of the furniture and the floor.

⑦ Yellow cabinets bring out the sunny aspects of the stone here – a pleasing contrast with the dark wooden table.

⑧ Go bold with a powerful turquoise hue to create a striking contemporary feel.

⑤

⑦

⑥

⑧

Hard Floors

Hard floors come in tile, brick or solid form and exhibit the durability that is crucial to good flooring. Interestingly, most hard flooring is created from the four elements: earth (the material), water (for binding), fire (for baking into shape) and air (for cooling). This elemental characteristic adds security, stability and a connection with the earth. Hard floors also do a fine job coping with the wear and tear of today's world.

Opposite: Porcelain tiles laid edge to edge add a touch of style. Porcelain has the ability to imitate stone finishes very well, but it also stands as a good-quality flooring material in its own right.

Below: Only an expert would spot that this floor is concrete, a material that is breaking free from its commercial and industrial usage because it can do a fantastic job – with amazing looks and great performance – in domestic settings.

People have been firing clay tiles for around five thousand years – and we haven't tired of them yet! Like stone, hard floors are made to last, offering excellent durability. This makes them the ideal material for rooms with traffic, such as entrance areas, hallways, kitchens and bathrooms, and messy environments such as utility rooms. However, hard floors have two major advantages over stone: lower cost and wider colour range.

Most hard flooring makes a smaller dent in the budget than stone. Perhaps more importantly, the wide range of hard flooring – and the huge variations in colour, pattern and texture – make it a very flexible component in the décor of a room. If you want a neutral setting to allow the rest of the room to shine, choose plain, regular tiles. If something really dramatic is required, play with patterns and colours until you get the right look for the room. You can also choose floor tiles to match or contrast with a backsplash or countertop tiles. This versatility allows you to create a unique look that blends the personalities of you and your home.

Most hard floors are made from tiles or bricks, which means they are laid in a pattern. Choosing the pattern can be as important as selecting the colour, since this can have a major impact on the mood of a room and how the eye is drawn. An overly complicated design makes the room seem cluttered, but a rigidly regular design can look uniform and clinical, so planning is crucial for this type of flooring.

Like stone floors, hard floors can be chilly and noisy. Although a cold floor can be an asset in some rooms (pantries, for example), other rooms may benefit from the addition of radiant heating and rugs. Hard flooring is also much more resistant to moisture.

Types and Finishes

Tiles are available in the many hues of ceramic or porcelain, or in the earthy quarry or terracotta tones. Finishes can be glazed or unglazed (an unglazed surface will require sealing). For the more adventurous, hard floors can also be made with glass or metal tiles. If a rustic, solid-looking base is required, perhaps leading out to the exterior of the house, clay bricks will work perfectly. In recent years the once expected dull, grey finish has been replaced with a host of distinctive top treatments, resulting in various colours and textures.

① Terracotta tiles are the classic choice for country-style properties.

② Porcelain is one of the most ancient high-quality materials, prized in China from the early dynastic period onwards.

③ For a stunning modern look, try glass for a genuinely different take on flooring.

④ Ceramic tiles are a popular choice because of the enormous range of colours, patterns and finishes available.

⑤ Concrete is the rising star of the flooring world because it forms a hard-working and attractive surface.

Ceramic Tiles

Ceramic tiles are made by moulding refined clay under great pressure and firing it at very hot temperatures. This creates a light but very hard tile with good resistance to water, stains and abrasion. These tiles are popular for bathrooms and are a good option for use on upper stories, especially in older houses where the weight of stone would be too much of a strain for the subfloor. Ceramic tiles are usually supplied glazed, so they do not require sealing (but the grout will).

① These chocolate-brown tiles complement the white paint and soft furnishings, as well as the wood panelling on the back walls. They do not compete with the eye-catching seats or the feature wall to the right.

② Tiles can replicate the look and feel of stone very well, but at a fraction of the cost.

③ Inlaid tiles enclose the eating area here without too harsh a transition from one area to another.

④ Laying tiles diagonally across the room adds liveliness and a sense of movement.

⑤ In a larger room, go for bigger tiles to coordinate with the room's proportions.

⑥ Use similar tiles on the walls and floor for a unified effect.

Ceramic tile is available in just about any colour, pattern and texture: you name it, you can find it. This flexibility makes it a designer's dream. Standard tiles are square (sizes vary), but other shapes – such as rectangles and octagons – are available and can be combined to make unique effects. In addition to the smooth, glazed finish, embossing or relief effects are also options. Consider installing strips of smaller tiles as a border or setting small diamond shapes in between tiles to create an inlaid pattern across the room.

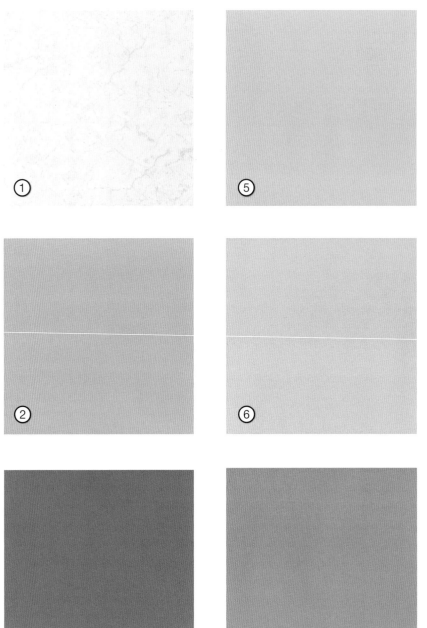

① Stone effect in white
② Smooth fawn
③ Smooth pastel mauve
④ Smooth dark turquoise
⑤ Smooth tan
⑥ Smooth grey
⑦ Smooth sage
⑧ Smooth dark aqua
⑨ Smooth navy blue
⑩ Smooth burgundy
⑪ Clouded grey
⑫ Slate effect in brown
⑬ Smooth mustard
⑭ Smooth malachite
⑮ Clouded tan
⑯ Slate effect in emerald
⑰ Smooth plum
⑱ Smooth ivory
⑲ Stone effect in ochre
⑳ Slate effect in charcoal

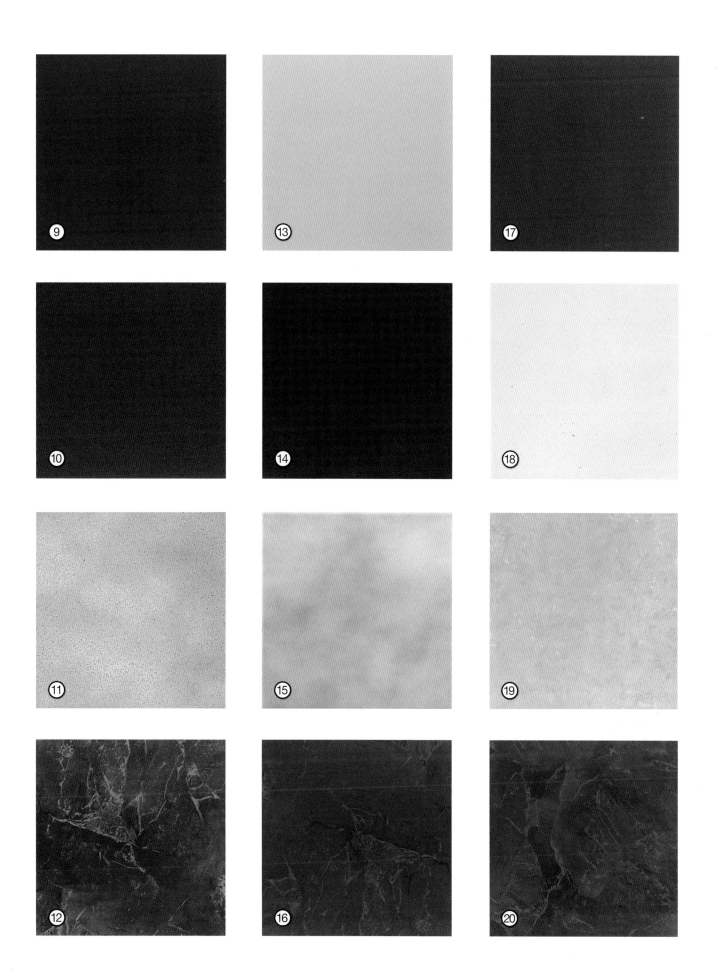

Continued from page 81.

① Marbled grey
② Marbled taupe
③ Slate effect in copper
④ Stone effect in ash
⑤ Marbled pearl
⑥ Slate effect in sand
⑦ Stone effect in rust
⑧ Stone effect in silver
⑨ Marbled ivory
⑩ Slate effect in black
⑪ Stone effect in cream
⑫ Slate effect in earth
⑬ Marbled alabaster
⑭ Slate effect in brushed black
⑮ Slate effect in aluminium
⑯ Stone effect in blizzard
⑰ Marbled powder blue
⑱ Clouded tan
⑲ Slate effect in olive
⑳ Stone effect in hazel

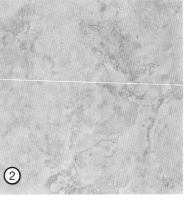

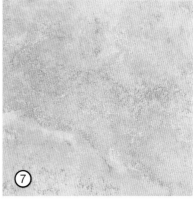

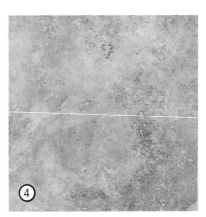

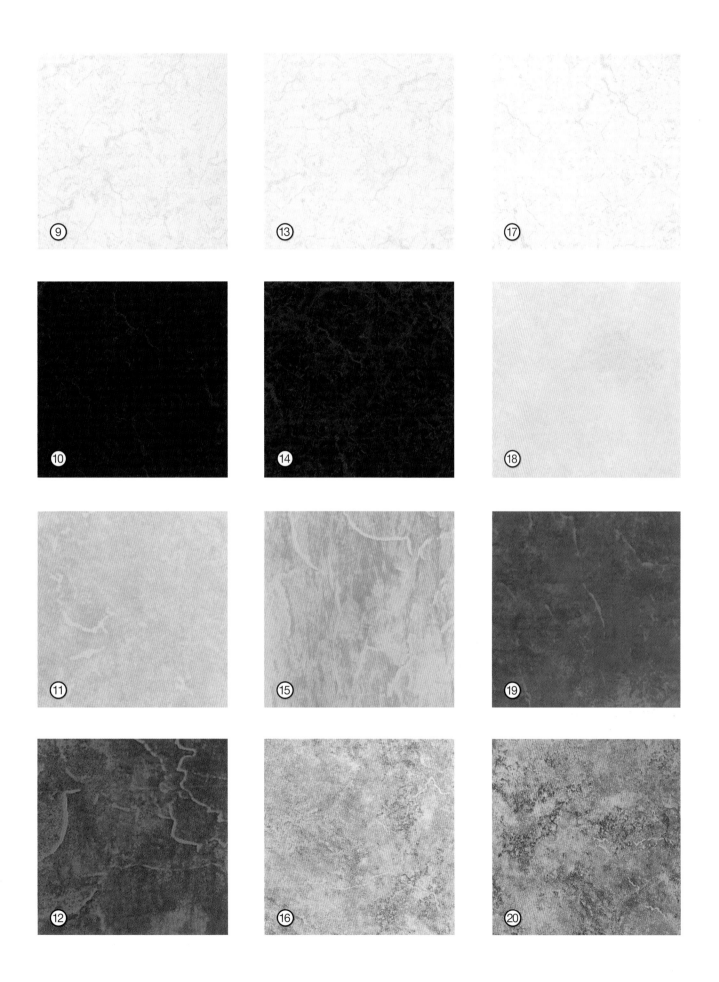

Continued from page 83.

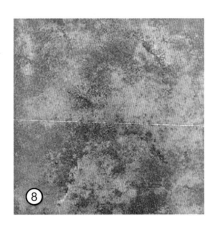

① Speckled blue
② Speckled evening sky
③ Speckled purple
④ Speckled royal blue
⑤ Speckled tan
⑥ Speckled brown
⑦ Speckled dark emerald
⑧ Rust effect brown
⑨ Rust effect blue
⑩ Sponged ivory
⑪ Sponged ocean
⑫ Sponged leaf
⑬ Sponged gold
⑭ Sponged brown
⑮ Sponged mist
⑯ Sponged silver birch
⑰ Sponged beige
⑱ Sponged bronze
⑲ Sponged steam
⑳ Sponged chalk

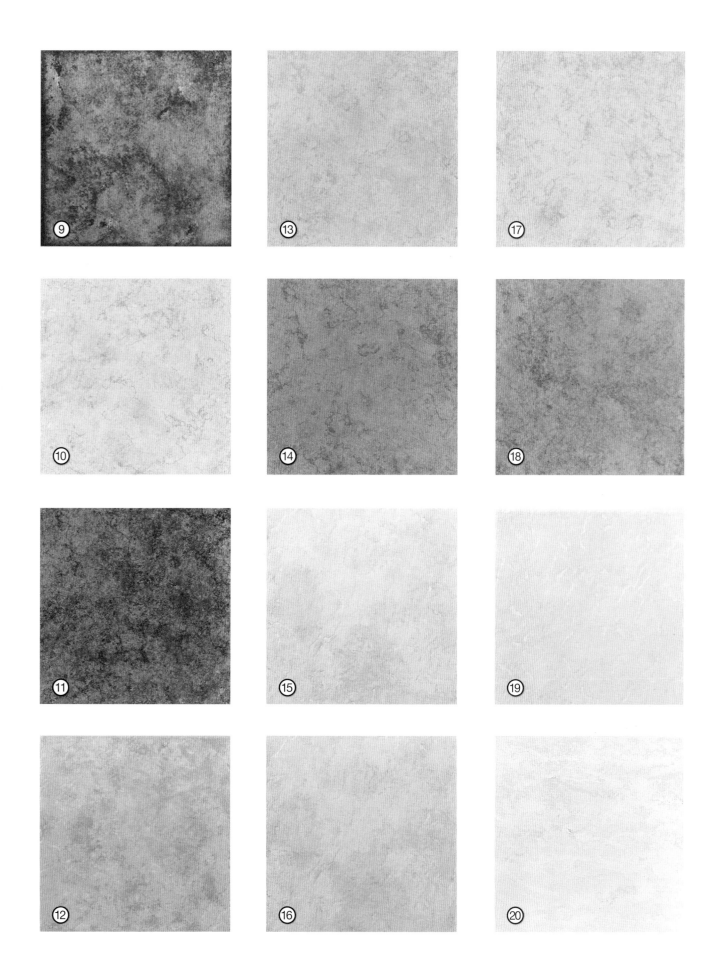

Concrete

Tell someone you're installing a concrete floor, and they will think that you've taken the whole utilitarian idea a little too far or have blown the budget and can't afford a 'real' floor. Show them a modern, decorative concrete floor, and they'll probably accuse you of lying, because concrete is no longer the dull grey material that is only fit for propping up the metal shelves in the workshop. It can be stained, painted and stamped to create an amazing array of colours and effects for your floor.

① Concrete is chameleon-like in its ability to mimic other materials, even shaped tiles.

② Concrete flooring can come in any colour and any design.

③ Stained and polished, this concrete surface looks like a marble floor.

④ Soften the harsh look of concrete with a mottled effect. This floor uses the greens that are also found in the furnishings to create an integrated design.

⑤ Beautiful warm tones echo the hue of the brick walls in this stylish kitchen.

Concrete can be coloured in several ways. The most effective way is to add pigment to the mixture (the same varieties that are added to plaster), meaning that the colour won't be affected by damage such as chips and scratches. Otherwise, there are special concrete-floor paints that can be applied to the existing concrete, with the advantage that you can add patterns and images to the floor. Another option is to apply acidic salts to the surface: these react with the calcium hydroxide in the concrete to create blues, greens and browns. The surface must be sealed to protect it from moisture and spills. Needless to say, installing a concrete floor requires a qualified professional.

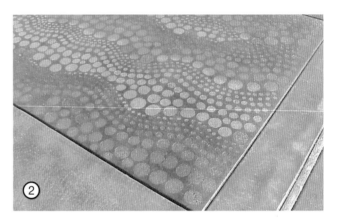

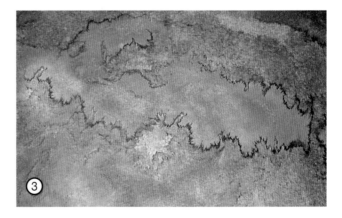

① Engraved mosaic
② Stencilled spots and curves
③ Acid-stained effect
④ Dyed abstract
⑤ Stencilled floral design
⑥ Engraved star
⑦ Dyed contours and shapes
⑧ Polished faux leather
⑨ Dyed kaleidoscope effect
⑩ Polished, mottled royal blue
⑪ Polished cloud design
⑫ Engraved faux brick

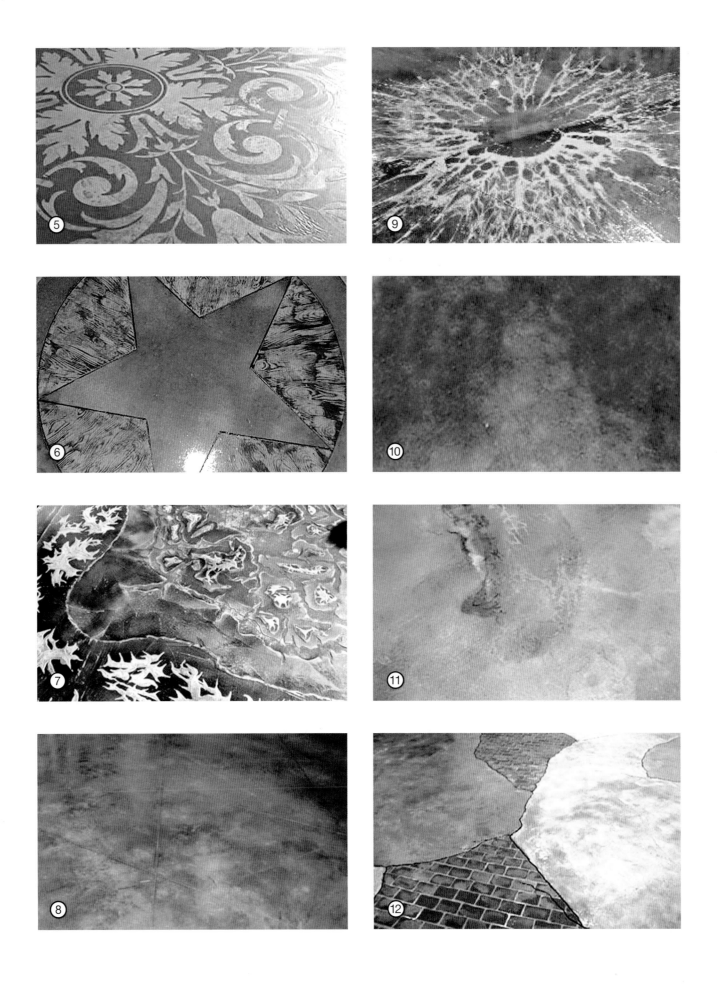

Clay Brick

Walls and floors have been made out of brick for thousands of years, and there is an old-world feel to a décor that features bricks, which is reassuringly warm and familiar. Flooring bricks (known as 'pavers') are more durable, lighter and thinner than construction bricks, so they don't raise the floor level. Brick is hard-wearing, slip-resistant – even when wet – and easy to maintain. Often identified with a rustic look, brick also works in contemporary settings because of its functionality and natural warmth. It is excellent for settings such as mud rooms and solariums that connect to the exterior. As with stone, a sturdy subfloor is vital to support the considerable weight.

① Brick is warmer than stone, so it can be well suited for living areas like this one, especially where it echoes the earthy colours of the décor.

② A herringbone brick floor matches the sloping terracotta roofing tiles in this unusual glass-ceiling kitchen, where modern meets rustic charm.

③ Brick flooring can even be suitable for bedrooms in a way that stone is not, because of its warm appearance and touch.

④ Brick looks best in strong natural light, which brings out the subtleties in its colours. There is no problem of glare because of its matte finish.

Brick finishes are usually a shade of earthy red, pink or brown. Paler, creamy effects can be achieved with calcium-silicate bricks (also known as sand-lime bricks), which are made by compressing mixed sand and lime under high-pressure steam. Pigments can also be added to produce a blue or green appearance. If you don't want uniformity, consider mixing coloured bricks in a checked or random pattern. The other important choice is about how bricks are laid: there are almost as many variations as there are for parquet (see pages 148–149). For small areas, go for the stack-bond pattern (even rows of bricks, resulting in straight mortar lines in both directions), but in larger spaces use basketweave or herringbone patterns to soften the look of the room.

① Plum mat
② Coral mat
③ Dusky pink mat
④ Warm beige mat
⑤ Beige mat
⑥ Pearl mat
⑦ Steel mat
⑧ Sea mist mat
⑨ Nut brown mat
⑩ Ruby mat
⑪ Copper mat
⑫ Bronze mat

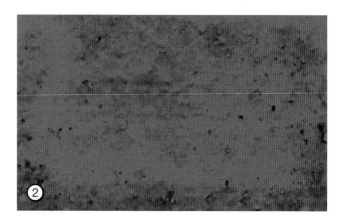

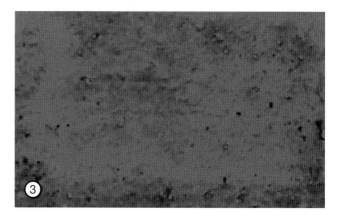

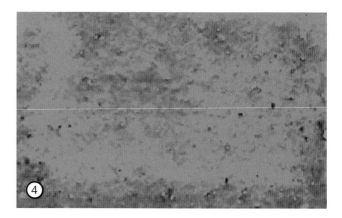

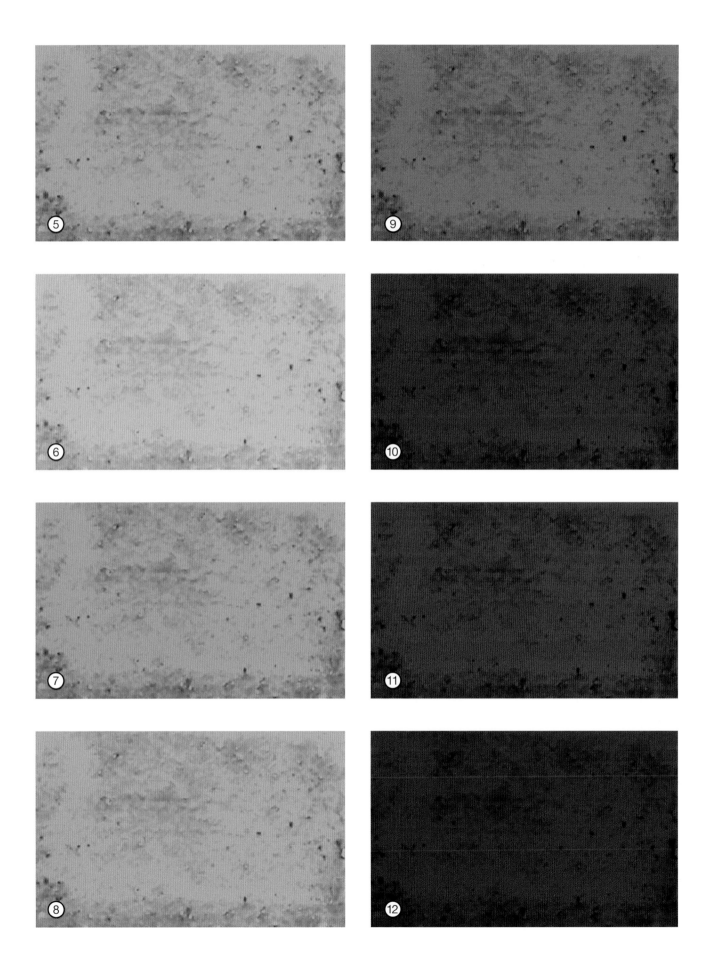

①

Glass

For the most spectacular and dramatic
floor, choose glass. The uncanny feeling of
walking on a transparent floor may not be
for everyone, but glass can be coloured and
sandblasted to create truly astonishing effects
that turn a floor into a work of art. Glass
has a sparkling quality that adds vibrancy
and lets in light, so it has practical benefits
in darker, enclosed settings. Glass flooring
is made from reinforced flat glass, usually in
square-foot panels, though many other sizes
are possible. Be aware that glass is heavy and
complicated to install – certainly not a job for
an amateur. Even if you don't like the idea
of a transparent walkway, mixing glass with
other flooring materials can create unique
and exciting results.

① A raised glass floor set over a side-lit beach display
exploits the unique qualities of glass, playing with light
and clear views

② A different material in this shower room would
accentuate its narrow confines, but glass transforms the
space, making it seem airy and interesting.

③ Glass introduces a vertical perspective to interiors,
completely transforming the way we experience
a room.

④ A glass floor requires carefully designed lighting and
colour choice. Underlighting and electric-blue paint have
turned this walkway into a work of art.

⑤ Two glass floors allow light to flood through this interior.
The uniformity of the grid pattern contrasts with the soft
curves of the staircase.

⑥ Glass mosaic tiles have a vibrant quality that injects
energy into a room, making this bathroom an interesting
and inviting space.

Glass tile (usually used as an accent, rather than to cover an entire floor surface) is available in numerous sizes, shapes and colours, while slip ratings and body strengths also vary. You will need to take advice on the right choice for your application. Few glass floors are made of clear glass: tinting the material adds to its appeal. The surface can also be etched with a pattern or motif. Large areas, such as walkways, are often sandblasted to reduce slipperiness and, on occasion, to conceal a vertiginous view.

① Translucent
② Stone-effect olive
③ Translucent ice blue
④ Translucent earth
⑤ Translucent terracotta
⑥ Stone-effect verdigris
⑦ Translucent royal blue
⑧ Translucent ocean
⑨ Translucent aqua
⑩ Stone-effect moss
⑪ Stone-effect ochre
⑫ Translucent grey
⑬ Stone-effect rust
⑭ Stone-effect foliage
⑮ Translucent mustard
⑯ Petroleum-effect pearl
⑰ Stone-effect blood-red
⑱ Stone-effect frost
⑲ Translucent rose
⑳ Petroleum-effect gold

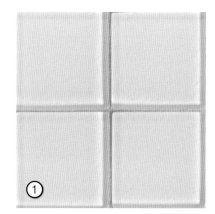

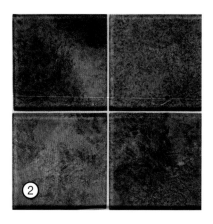

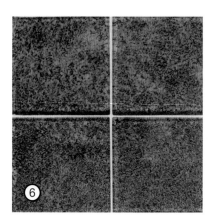

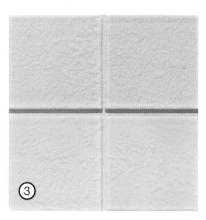

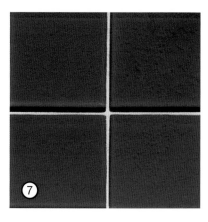

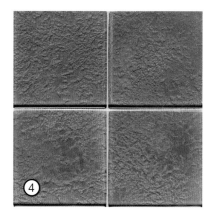

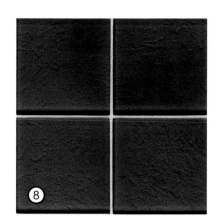

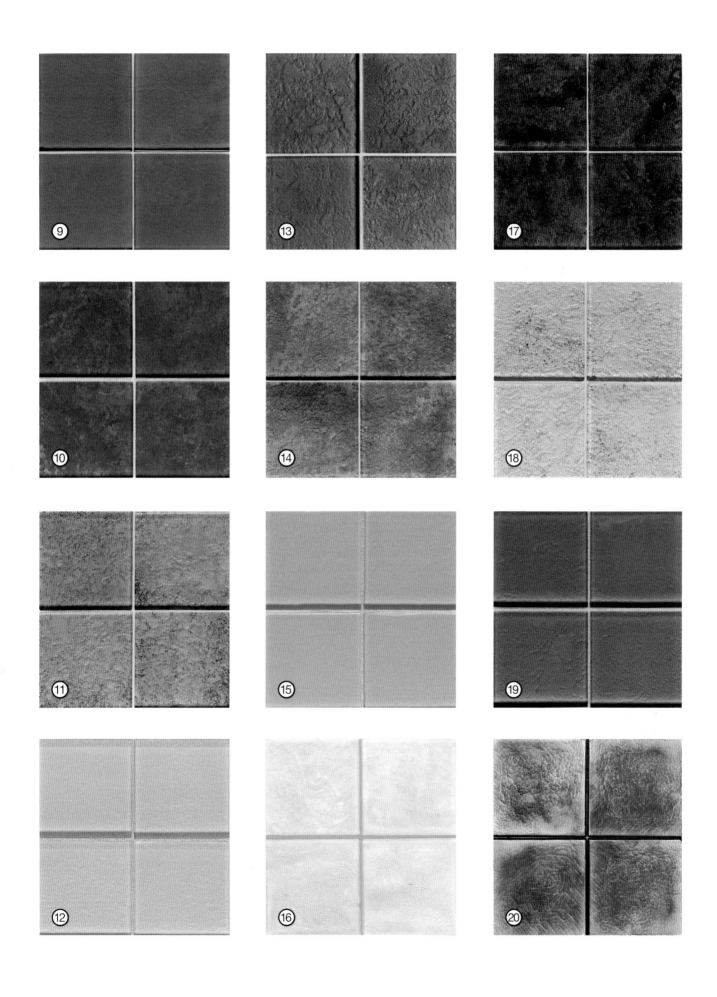

Continued from page 97.

① Etched tile in cream
② Etched tile in ice
③ Etched tile in jade
④ Etched tile in apricot
⑤ Etched tile in ivory
⑥ Etched tile in cherry
⑦ Etched tile in cerulean
⑧ Etched tile in black
⑨ Etched tile in sea green
⑩ Etched tile in silver
⑪ Etched tile in pink pearl
⑫ Etched tile in ultramarine
⑬ Etched tile in mint
⑭ Etched tile in sapphire
⑮ Etched tile in petroleum black
⑯ Etched tile in blue pearl
⑰ Etched tile in purple
⑱ Etched tile in chocolate
⑲ Etched tile in blue-green
⑳ Etched tile in black pearl

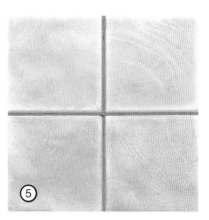
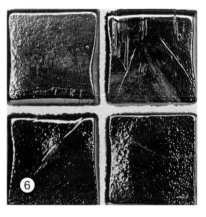
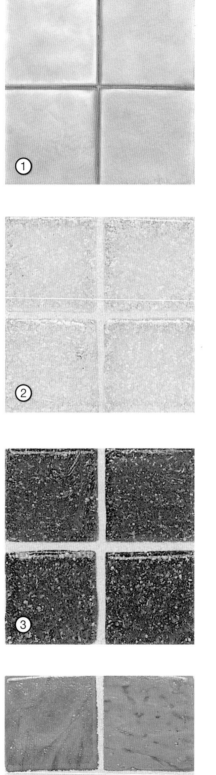

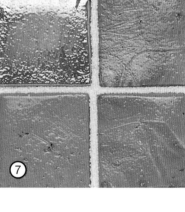
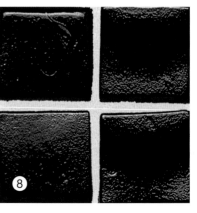

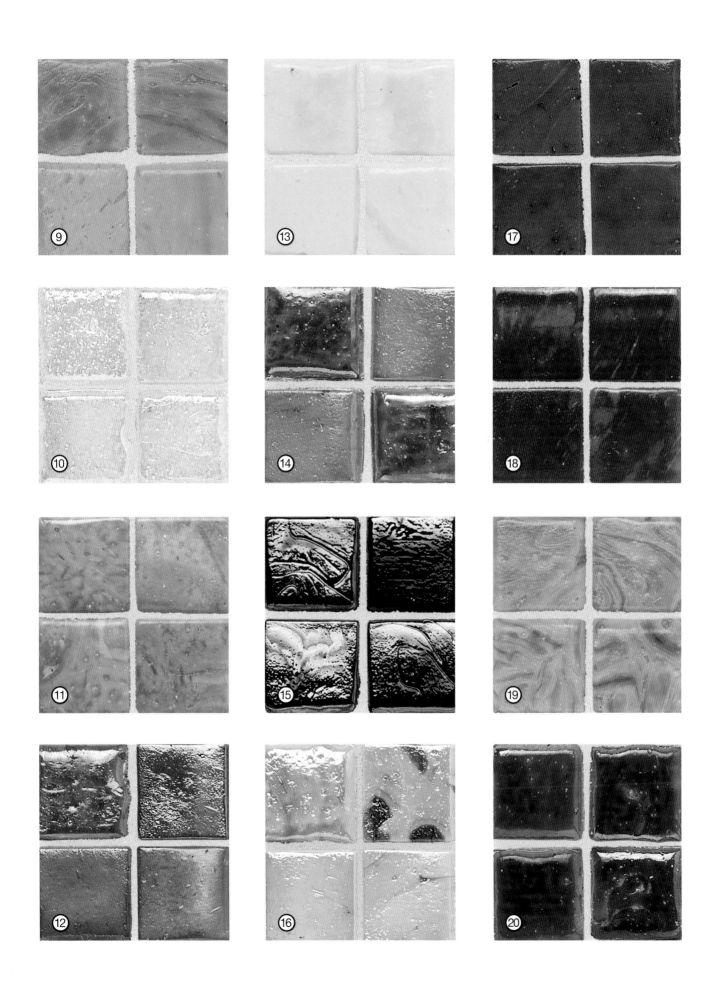

Metal

A metal floor provides the ultimate contemporary look: sleek, sophisticated and wonderfully functional. In addition to offering a surprisingly wide range of colours, this material will also contribute to your decorative scheme by reflecting light and colour in its distinctive shimmer. There are drawbacks: metal is hard, noisy and cold and requires expert installation to prevent flexion. It certainly doesn't suit every room style, but nothing can beat it for a sleek, industrial look. Metal shines in bathrooms and, especially, utilitarian kitchens. If the texture is slippery, a nonslip finish can be burnished onto the surface.

① In the kitchen you can match a metal floor with steel appliances and equipment, such as the stove, sink and even the electrical outlet.

② This contemporary setting features an unusual combination of metal, wood and glass flooring.

③ As with wood, inlays can be set in metal to add colour and create visual interest.

④ Metal and wood complement each other in this kitchen.

⑤ This unusual finish is created with brushed metal planks and coloured insets.

Metal flooring is made from aluminium, galvanised steel, or sometimes bronze – often recycled from other uses, so it is environmentally sound – but it can be treated to look like brass, copper, bronze and many other metals. The illustrations on these pages are of metal in sheet form, but it is also available in tiles (see pages 104–105). Metal can be processed into a satin, mirror, lacquered or matte finish and it can take on a whole new look if exposed to water or other elements, which may or may not be desirable – check this with your supplier.

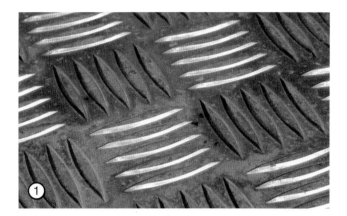

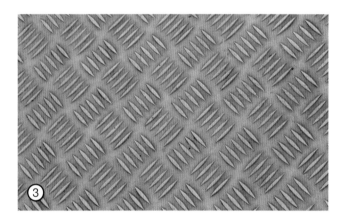

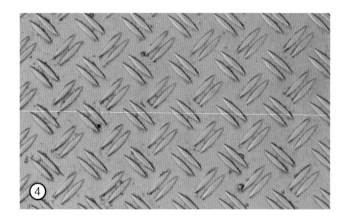

① Five-line relief, mat
② Five-line relief, polished
③ Five-line relief, green
④ Two-line relief
⑤ Two-line relief, bronze
⑥ Crossing design, polished
⑦ Crossing design, mat
⑧ Crossing design, blue
⑨ Oval holes
⑩ Round holes
⑪ Grill
⑫ Studded

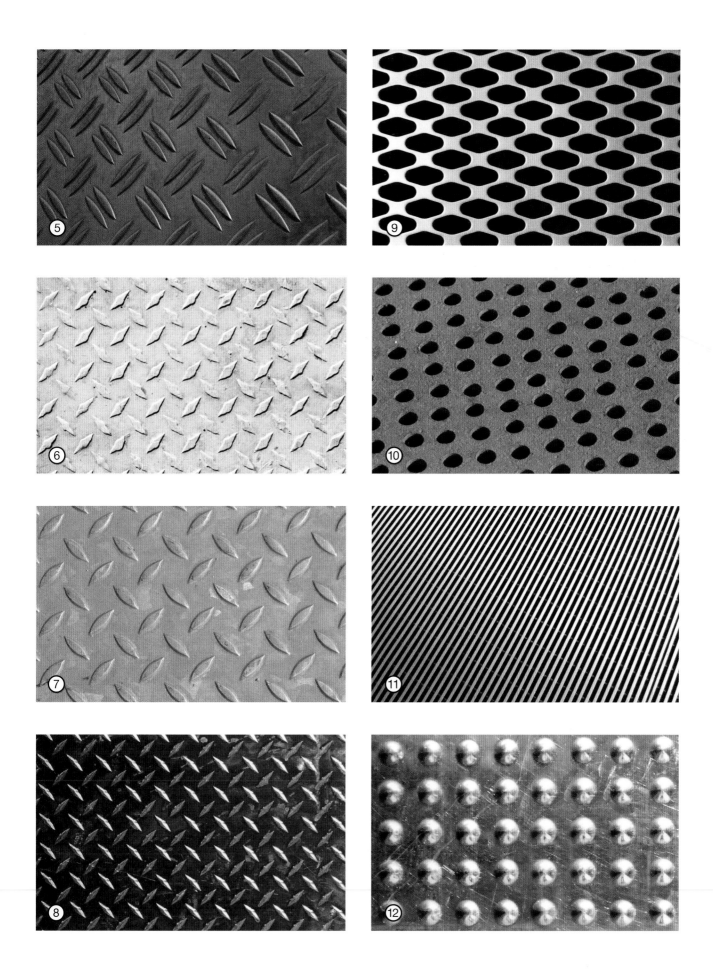

Metal tiles offer greater design and installation flexibility than sheet metal, as well as the possibility of matching the flooring material with the finish on the walls – either throughout or with the use of accent tiles. Most flooring tiles have a matte glaze for better wear and stain-resistance. Some particularly upmarket tiles are made of solid aluminium and decorated by hand, creating truly unique finishes.

① Satin matte in silver
② Satin matte in blue-grey
③ Satin matte in light bronze
④ Satin matte in amber copper
⑤ Satin matte in warm grey
⑥ Satin matte in bright gold
⑦ Satin matte in dark bronze
⑧ Satin matte in new copper
⑨ Satin matte in rust
⑩ Geometric pattern, silver squares
⑪ Floral design in alpha blue
⑫ Cross design, camilla
⑬ Satin matte in mink
⑭ Geometric pattern, grey crosses
⑮ Abstract design in aqua
⑯ Natural design, filigree
⑰ Midnight satin matte
⑱ Geometric pattern, grey triangles
⑲ Cross design, black flowers
⑳ Swirls design, black

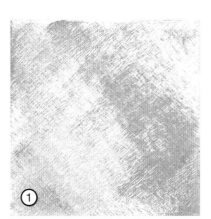

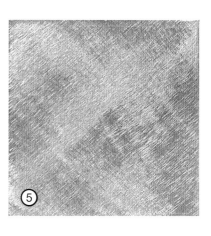

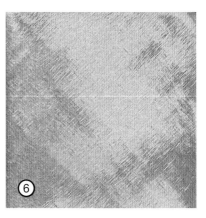

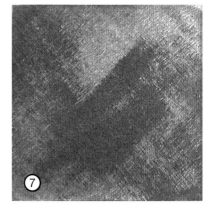

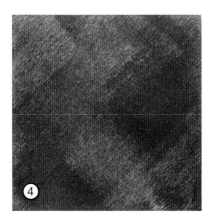

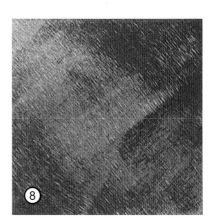

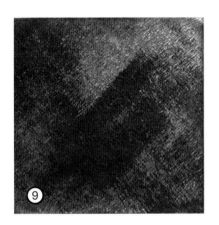

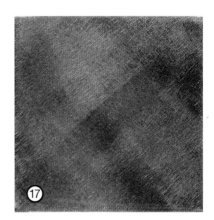

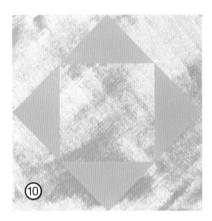

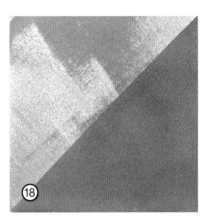

Quarry Tiles

Quarry tiles were developed during the Victorian era as a mass-produced alternative to the handmade terracotta tiles that were so popular in kitchens and passageways. Made by extruding unrefined clay or shale into a mould before the material is pressed and heated, quarry tiles look very similar to the terracotta variety, but they are more durable and water resistant. This makes them better suited for kitchen applications – although they are much colder to the touch. Also known as 'quarries', these tiles are rougher than ceramic tiles and as a result have good slip-resistance.

① Quarry tiles look similar to brick when set wide apart in this rustic-style bedroom.

② With its excellent water-resistant properties, this material is superb for creating a warm and unique bathroom.

③ Mix the colours available to create a pleasing geometric effect.

④ Stone works works well alongside stone, as in the interesting variety of finishes here.

⑤ Sealed and polished to a shiny finish, quarry tiles combine warmth and style with echoes of their origins in the Victorian period.

Quarry tiles, generally identified with earth tones, vary widely in colour and finish. They are available in browns, reds and even shades of greens and greys. Some glossy and vivid finishes are also available. However, the bulk of these tiles come in similar shades to those of terracotta. Generally available in standard squares, quarry tiles take on an especially rustic look when they are separated by wide gaps. They can look much more stylish, though, when varying sizes are set in a formal pattern.

① Earth, with traction finish
② Terracotta, smooth finish
③ Silver, smooth finish
④ Square beige, smooth finish
⑤ Grey, with traction finish
⑥ Steel, smooth finish
⑦ Plum, smooth finish
⑧ Tan, smooth finish
⑨ Copper, smooth finish
⑩ Square black and red, smooth finish
⑪ Textured earth
⑫ Light grey, smooth finish
⑬ Fawn, smooth finish
⑭ Rectangular charcoal-streaked
⑮ Textured grey
⑯ Dark grey, smooth finish
⑰ Slate, smooth finish
⑱ Rectangular wood-streaked
⑲ Textured beige
⑳ Light blue, smooth finish

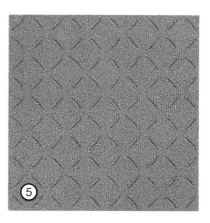

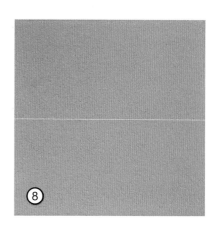

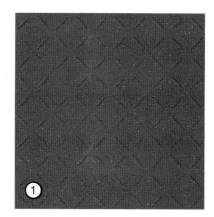

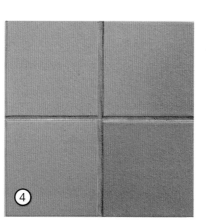

9

13

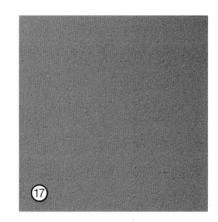
17

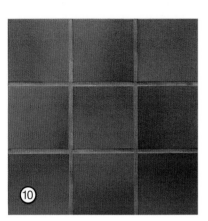
10

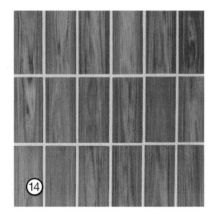
14

18

11

15

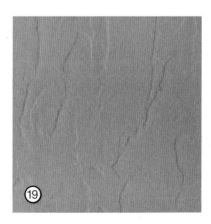
19

12

16

20

Porcelain

Porcelain tiles (also known as 'pavers') are similar to ceramic tiles but of a higher quality (and price). This is because they are made from fine, white clay that is fired at very high temperatures. The superior quality of the material and the particular firing process produces a very dense, hard tile that is waterproof and stain-resistant, making it a great choice for high-traffic areas. The irregularities caused by firing tiles at high heat may prove to be problematic if you plan to install tiles yourself – distortion will make it more difficult to achieve a level floor surface.

① Porcelain can resemble stone, such as limestone, and is guaranteed to add a sense of luxury and richness to the décor.

② This unusual porcelain floor has a metallic finish created by adding zinc oxide or bronzed copper to the surface of the tiles.

③ Porcelain has traditionally been used in ornate settings where an extravagant effect is desired.

④ Laying tiles of various sizes in a formal pattern creates an elegant effect that is dignified without seeming regimented.

⑤ Porcelain is a distinguished material that adds an air of refinement to many settings. Although it is the most expensive of the fired tiles, it costs less than the stone that it can mimic so impressively.

Porcelain has a translucent quality that has made it a popular material for figurines and other decorative objects. When used on floor tiles it provides rooms with a distinctive luster. Porcelain comes in a wide variety of glazes and finishes, ranging from matte to high gloss. It can resemble stone and even glass, and you can choose from an enormous palette of pastel colours. The tiles are often polished, but the surface can be embossed with a raised grip to improve slip resistance.

① Smooth ivory
② Sponged beige
③ Stone-effect charcoal
④ Marbled cream
⑤ Veined cream
⑥ Stone-effect tan
⑦ Stone-effect olive
⑧ Marbled grey
⑨ Smooth blue
⑩ Smooth grey
⑪ Flecked blue
⑫ Stone-effect grain
⑬ Smooth sage
⑭ Smooth azure
⑮ Stone-effect sand
⑯ Stone-effect ash
⑰ Smooth mustard
⑱ Smooth cobalt blue
⑲ Stone-effect heather
⑳ Stone-effect red sand

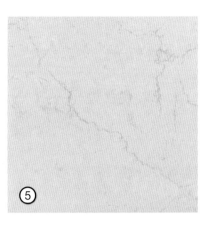

①

⑤

②

⑥

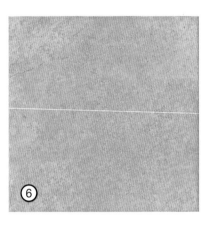

③

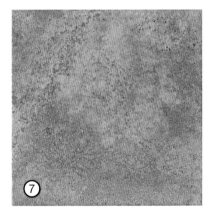

⑦

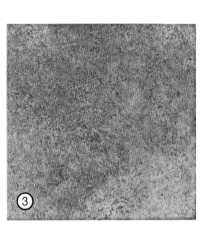

④

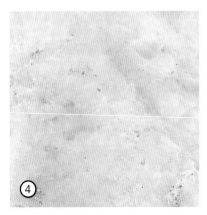

⑧

Continued from page 113.

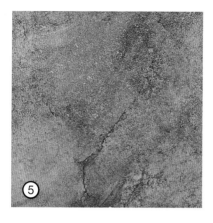

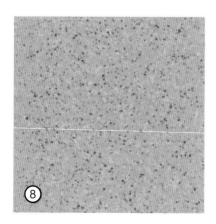

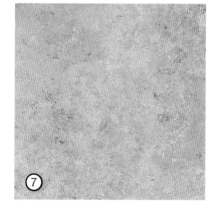

① Stone-effect lead
② Stone-effect blue-grey
③ Mottled cream
④ Smooth grey
⑤ Stone-effect pepper
⑥ Stone-effect pearl
⑦ Mottled beige
⑧ Flecked grey
⑨ Sponged grey
⑩ Sponged auburn
⑪ Stone-effect verdigris
⑫ Mottled tan
⑬ Sponged gold
⑭ Stone-effect rust and grey
⑮ Flecked red
⑯ Veined iron grey
⑰ Brushed henna
⑱ Stone-effect bronze
⑲ Mottled grey
⑳ Veined tan and brown

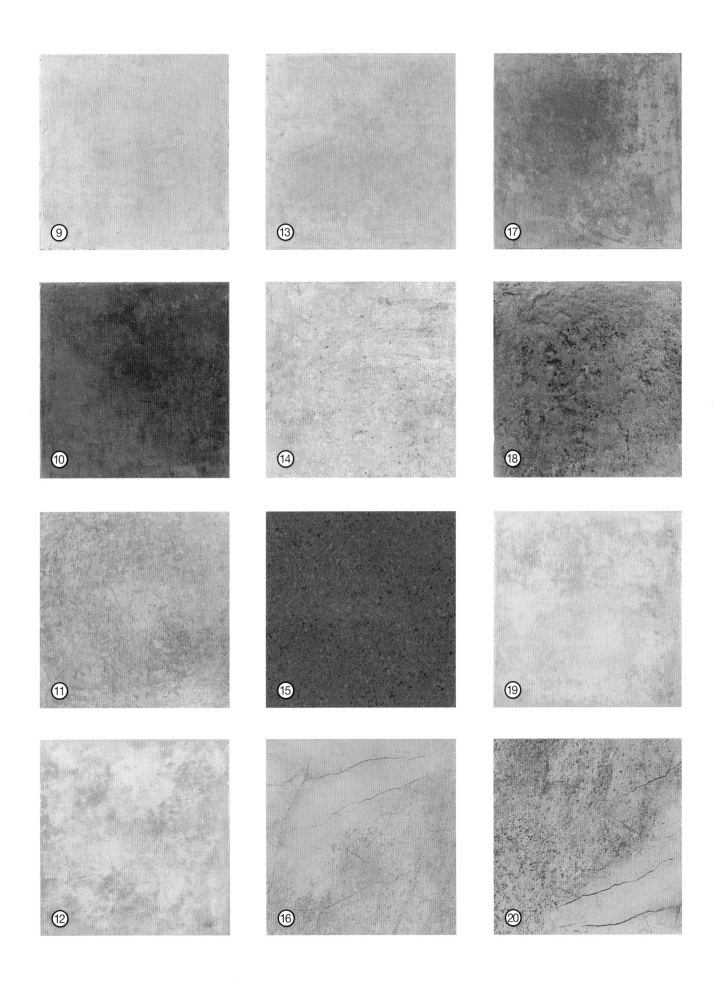

Continued from page 115.

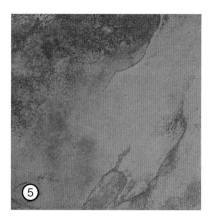

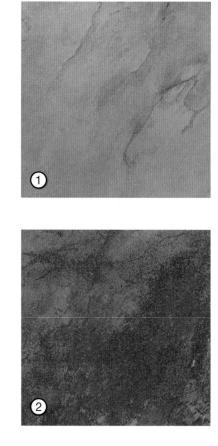

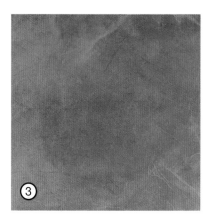

① Stone-effect iron and olive
② Stone-effect plum and tan
③ Slate-effect grey-blue
④ Slate-effect sorrel
⑤ Stone-effect ecru
⑥ Stone-effect seashell
⑦ Slate-effect blue
⑧ Slate-effect deep blue
⑨ Streaked tan
⑩ Smooth taupe
⑪ Stone-effect mist grey
⑫ Stone-effect soft grey and olive
⑬ Streaked wheat
⑭ Flecked tan
⑮ Granite-effect gravel
⑯ Granite-effect mottled brown
⑰ Smooth grey
⑱ Flecked straw
⑲ Granite-effect pebble
⑳ Granite-effect blue and brown

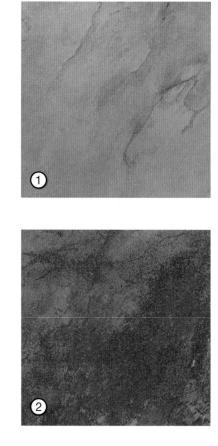

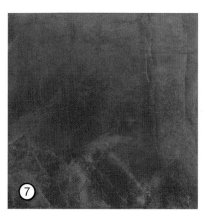

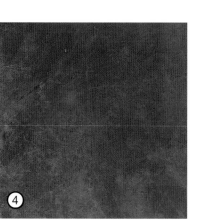

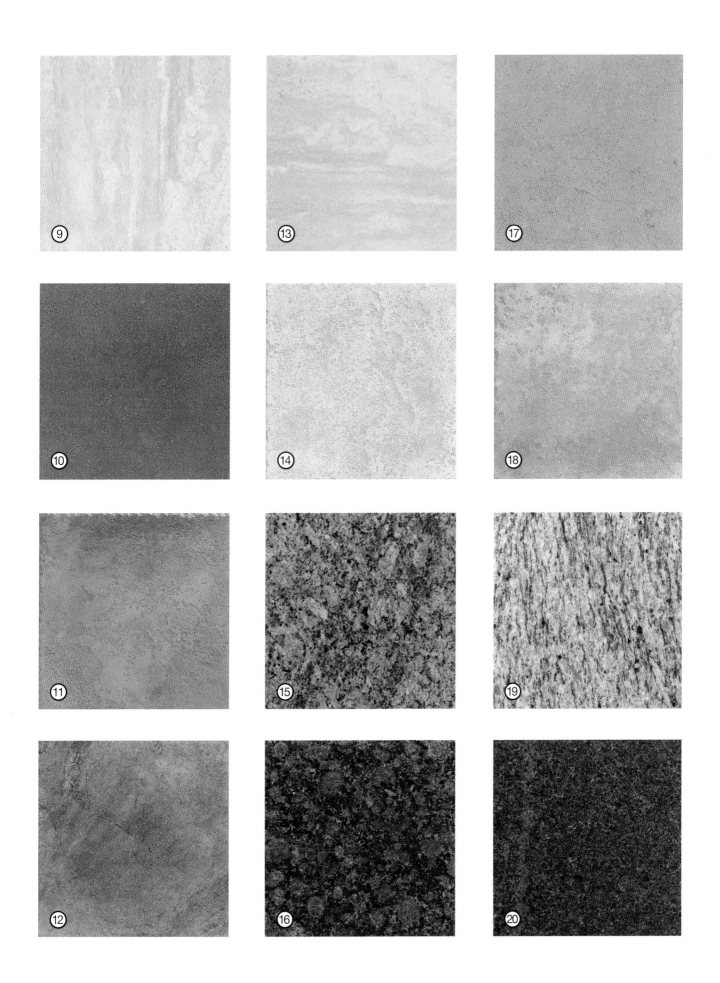

Terracotta

Terracotta is the warmest of all tiles, emanating a mellow charm. Like quarry tiles, it is formed from clay, but is fired at much lower temperatures. This process leaves tiny air pockets in the material, resulting in a porous surface that retains heat well. This also means that it can easily be damaged by water if not properly sealed. It also explains the differences in colour tone between tiles in the same batch, which adds to their rugged individuality. Terracotta is one of the oldest flooring materials in the world, and there is a thriving market in antique tiles. Modern tiles can be artificially distressed to simulate centuries of wear – or you can try laying the tiles upside down, which will also give them an aged look. These tiles come in a range of shapes, including squares, rectangles, hexagons and octagons.

① Lay the tiles at an angle to the walls to create a less formal effect.

② The natural, down-to–earth character of terracotta makes it especially suitable for use in kitchens.

③ Wide grout lines emphasise the rustic character of this ancient tile.

④ The many small variations in tone that are typical of terracotta have a positive effect in large rooms, adding interest and character.

⑤ Terracotta works well with pale woods and brick walls.

Terracotta tiles come in a range of earth colours, from brick-red to soft pink and includes fiery oranges, mustard yellows and an array of dusky browns. Glazed variations bring in a whole new palette of blues and greens. Tones vary within batches of these tiles, so first lay them out on the floor to see how you can either disguise or accentuate these inevitable differences in your laying pattern. Handmade tiles will also vary slightly in thickness, requiring extra attention during installation to keep the floor level.

① Henna
② Soft earth
③ Hazel
④ Mottled brown
⑤ Copper
⑥ Cinnamon
⑦ Russet
⑧ Tan
⑨ Rustic
⑩ Handmade
⑪ Geometric pattern
⑫ Smooth
⑬ Polished
⑭ Close-fitting
⑮ Rustic geometric pattern
⑯ Chiselled

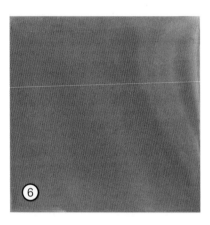

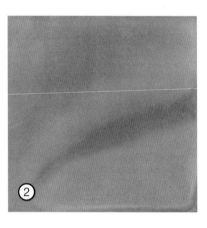

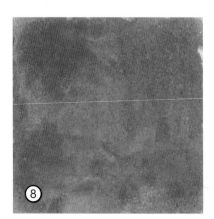

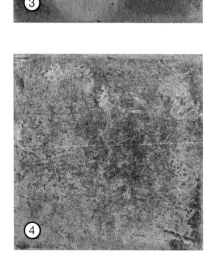

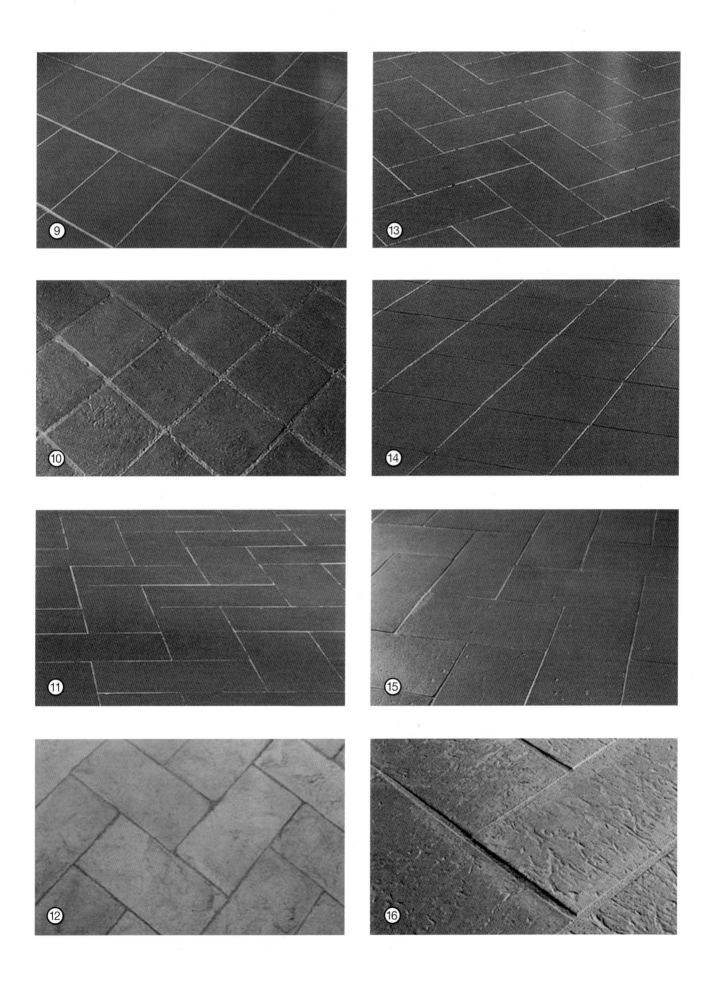

Combining Tiles To Make Patterns

Once you have chosen your tiles – or, even better, as you are choosing them – you need to decide how they will be laid. First, confirm the shape: most tiles are squares, but rectangles (which have a Hispanic flavour), octagons, diamonds and triangles are possibilities. Second, are the tiles best laid edge to edge or with a grouted gap – and how wide should this gap be? The wider the gap, the more rustic the look. Finally, think about the pattern: the smaller the room, the simpler your design should be. The following illustrations show some of the options available.

Standard draughtboard

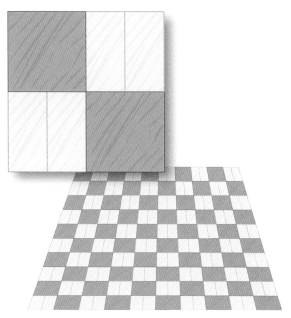

Alternating checkerboard

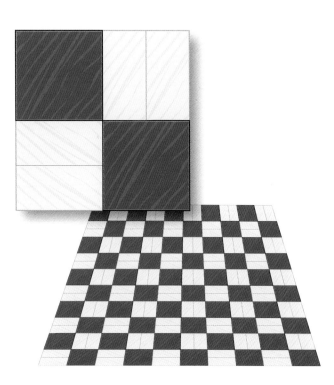

Alternating draughtboard variation

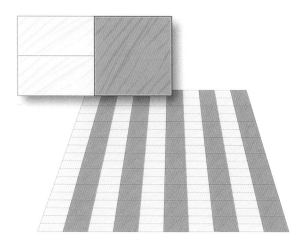

Alternating stripe

Basketweave

Basketweave variation

Diagonal stripe

Barred square

Random block pattern

Diagonal block pattern

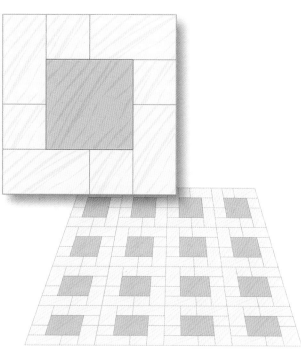

Block pattern with random square border

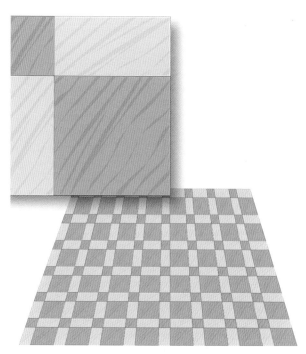

Block pattern with inlaid block border

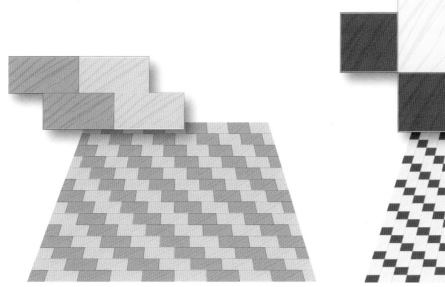

Running bond with alternating brick colours

Diagonal pattern

Pinwheel

Diagonal block pattern variation

Herringbone

Double herringbone

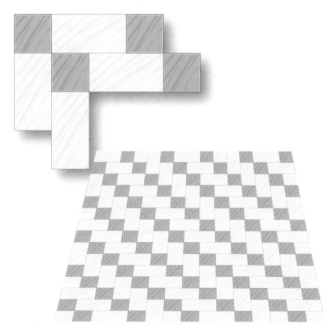

Diamond herringbone

Railroad bond

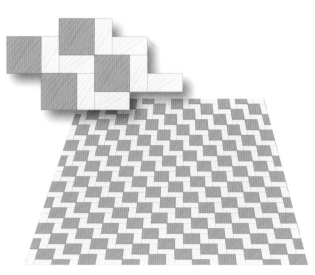

Diagonal block variation

Diagonal variation

Spiral

Interlocking spiral

Using Colour

The traditional hard-floor colours are the dusty reds and browns of terracotta or quarry tiles, but the options are so much wider. For example, concrete can take on any colour, while ceramic tiles come in any pattern or hue. Tiles allow the introduction of patterns from the grid of grouting. This adds texture and allows you to play with adding a stronger colour than you might otherwise consider. If you stick with a neutral, like the grey opposite, you can afford to be bold with your colour choices.

① A neutral setting with grey tiles is lifted by a couple of accent colours.

② The dusty red of a bright terracotta floor feels even more Mediterranean when paired with yellow and olive.

③ We identify soft blue with sophistication. Here it is paired with one of its most effective contrast colours, yellow, to bring some welcoming vivacity.

④ Floors set moods: earthy brown with orange and cream has a cool, 1970s retro feel.

⑤ Keep the walls natural when you go bold elsewhere, as with this striking pink-and-black draughtboard tile floor, with violet and red accents.

(6)

(8)

(6) Pink doors and yellow walls create a smart and vibrant feel as the eye is drawn along the room towards the bright colours.

(7) A bold purple teamed with dark wood creates a more sophisticated, modern look.

(8) In this hallway natural tones in light brown and grey create a calm ambience.

(9) Red and black is a lively combination that works well with neutrals and makes the doorway very inviting.

(7)

(9)

Wood Floors

Wood flooring has long been a favourite flooring choice, and for several reasons. First, wood is beautiful, adding warmth and affinity to nature. Secondly, wood offers the performance and style of stone at a more affordable price. Thirdly, well-sourced wood meets the rising demand for environmentally responsible products. Finally, easy-to-install, less costly floors have been developed, such as prefinished, engineered, and glueless varieties.

Opposite: Wood looks good with more wood. Visually unifying this large room, the merbau floor contrasts beautifully against the bamboo uprights that screen the seating area.

Because it is a natural product, wood must be harvested sustainably so that future generations can enjoy it. Salvaged from its first use in old barns or as railway sleepers, reclaimed wood – which tells the story of a long and productive life through minor blemishes, such as dents – provides another eco-friendly alternative.

Wood enhances almost any decorative scheme – it's simply a matter of finding the right kind. From pale maple to rich, dark walnut, wood offers all the colours of the forest. The heavy grain of hardwoods, such as oak, is great for a country style, but for a contemporary look, try using thin planks of pale beech. Once installed, wood can be sealed clear or stained to bring out the natural elegance of the grain and the characterful knots. Varying the types of wood or the choice of stain or clear seal, creates an infinite number of pleasing combinations and the surface can be sanded and refinished as the flooring wears.

Wood is durable enough to cope with the constant traffic of everyday life, so it will provide reliable flooring for any room in the house. However, wood and water do not mix, so wood requires regular sealing if it is installed in a damp environment such as a bathroom. Warmth and 'give' make wood much more comfortable to stand on for long periods than stone, and it also offers reasonable sound insulation, though not as good as carpet. In kitchens it can deal with spills, but a dropped knife (or a spiked heel) will make a carved niche in your floor, so beware or accept these nicks as part of the patina of life. The densest hardwoods are the most durable, but softwoods are also fine if you intend to paint the end result. The availability of home-grown timber and imported exotic varieties makes for a wide choice of finishes.

Below: Dark oak is proving to be popular for use in kitchens as its look combines warmth and style. These planks have been laid along the room to accentuate its length.

Types, Sizes and Finishes

Traditional solid-wood flooring comes as strips (narrow boards), planks (larger, random lengths and widths, usually 75, 125 and 175 mm wide (3, 5 and 7 inches), or parquet, where short pieces are arranged in patterns and usually glued into a 12-inch (30-cm) tile. A less expensive option is engineered wood, which has a thin hardwood veneer. Installation has never been easier with floating tongue-and-groove systems and the even easier 'click' method.

① The wengé plank flooring shown here demonstrates the dense finish of this dark African wood, which is also used to make masks and musical instruments.

② Walnut has a distinctive grain that adds texture to this large room. Rugs are used to define the eating and seating areas.

③ Bleaching has created a dramatic contrast of the dark and pale colours within each board of wood in this ash floor.

④ Teak has a certain warmth that allows the walls and furniture here to be finished in cooler, neutral colours.

⑤ Laying the floor in a herringbone pattern stops the eye from being drawn in one direction too much, which prevents an overly clinical effect.

⑥ This unusual flooring combination of wood and stone is effective because the intricacy of the parquet pattern echoes the density of the stone finish.

Wood Choices

Every piece of wood is different, and you have a choice of more than 50 domestic and exotic tree species for your floors. Your choice will be influenced by the colours and textures that are right for the room's décor – but remember, wood finishes will get darker with age.

Most floors are hardwood, which is much more durable than softwoods such as pine and fir. The most common hardwoods used in flooring are ash, beech, maple and oak. Bamboo is also increasingly used now because of its environmental advantages (see page 138). Consumer demand for sustainable products is hugely influencing the flooring industry. Because of this demand, several companies in the U.S.A. and Canada are now harvesting hardwood species responsibly. When buying imported tropical wood, look for certification by the Forest Stewardship Council, which encourages good management of forestry.

① Rooms with plenty of natural light can take a dark floor. The rich, deep hues of this cherry wood are echoed by the exposed wooden beams in the ceiling.

② Oak creates a very warm, welcoming feeling that balances the austerity of this contemporary room.

③ This rosewood floor provides a genial contrast to the cool white of the walls.

④ The deep, richly textured finish of jatobá is a good counterpoint for the blue-grey stone walls in this striking room.

⑤ Maple is a popular choice when a lighter tone is required. Here, it provides an effective contrast to the dark wood of the furniture.

⑤

④

Acacia grows in Africa, India, Australia and Hawaii. It is known for its dark, rich colours and amazing hardness – it is much tougher than oak. This means it resists warping and other damage much better than almost any other finished wood.

Andiroba is a hard, durable wood from South America. It is reddish-brown in colour, with a straight, sometimes interlocked grain and quite a coarse texture.

Ash and beech are both grown in the central and eastern states of America. They are light-toned woods that are well suited to a contemporary setting. Ash is a relatively new flooring option. The grain is usually relatively straight, with an occasional wavy line.

Beech has a tight and straight grain, giving a neat, even texture. This wood is extremely strong, so it is especially useful for flooring applications where traffic will be heavy. It is also often used in block flooring (see page 148). American beech has a coarser appearance than its European counterpart.

① Acacia
② Andiroba
③ Pale ash
④ Medium ash
⑤ Straight-grain ash
⑥ Dark ash
⑦ White ash
⑧ Varied-grain ash
⑨ Wavy-grain beech
⑩ Varied-grain beech
⑪ Pale beech
⑫ Rich beech

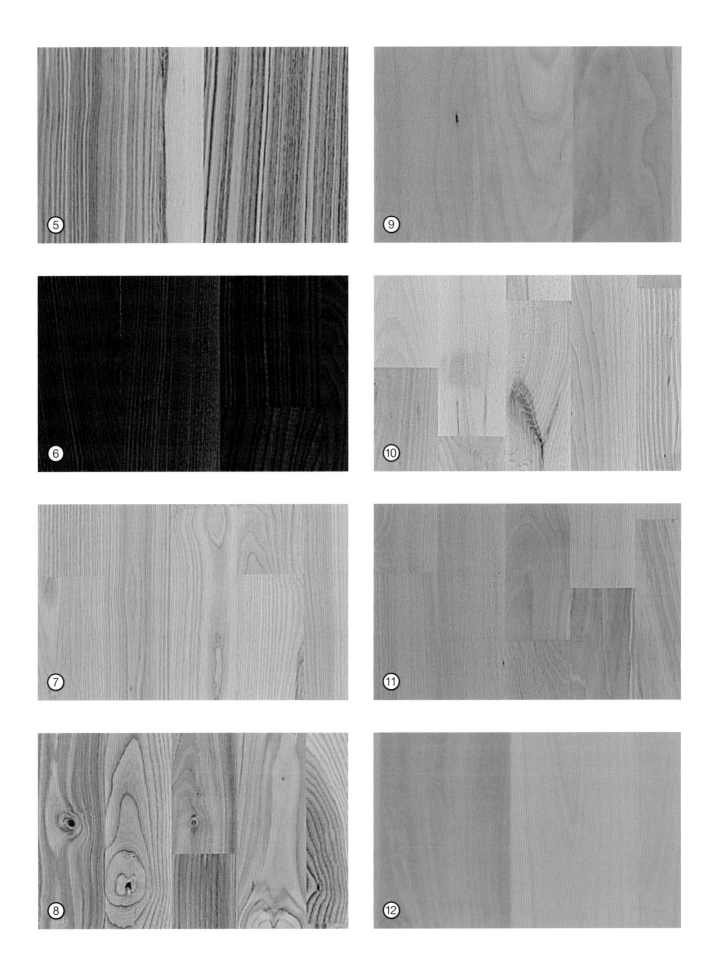

Bamboo is increasing in popularity and is rivalling maple as the favourite for a light-coloured finish, partially because it can be harvested sustainably. Bamboo is actually a grass, strips of which are treated to create an elegant-looking, high-performance wood that can be stained to match any décor. Slicing it on a flatter angle produces patches in the grain.

Birch is finely textured with colours varying from pale cream to light tan, sometimes with a hint of reddish-brown within its straight grain. The pale finish of several varieties is popular in contemporary rooms, but some people find it lacking in character.

Amboyna burl is a rare, exotic hardwood that comes from the Narra Tree which grows in southeast Asia.

Cherry wood has a warm, reddish finish with a slightly wavy, fine grain and an even texture that adds richness to a room. American black cherry comes from Canada and the U.S.A., but there is also a Brazilian variety, known as jatobá or Brazilian cherry (see page 140).

① Flat-grain, natural bamboo
② Flat-grain, caramelised bamboo
③ Vertical-grain, natural bamboo
④ Vertical-grain, caramelised bamboo
⑤ Grey birch
⑥ Medium birch
⑦ Pale birch
⑧ Contrasting-grain birch
⑨ Amboyna burl
⑩ Wavy-grain cherry
⑪ Straight-grain cherry
⑫ Characterful cherry

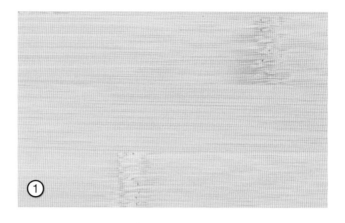

①

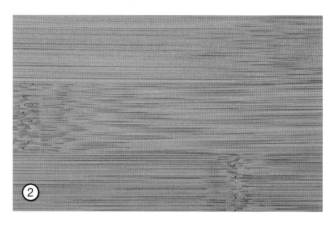

②

③

④

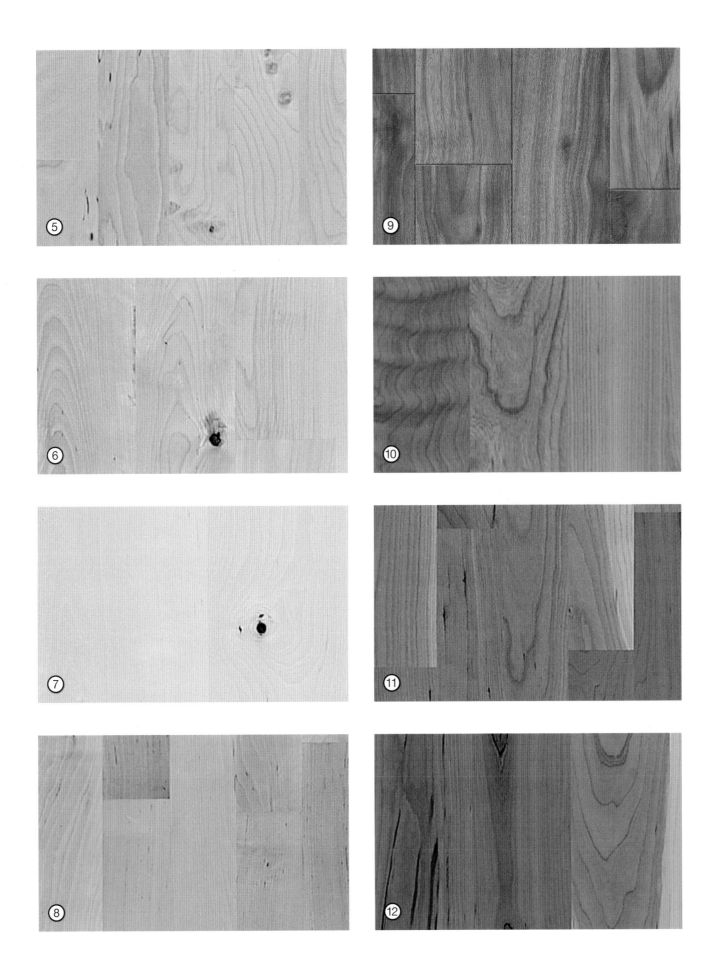

Hickory is a robust wood, making it an ideal flooring material. Higher-quality hickory is cut from sapwood, which is lighter than the reddish-brown heartwood.

The exotic woods ipê, jatobá, jarrah and santos mahogany are excellent for commercial and high-traffic environments, though the cost of importation raises the price.

Ipê is often sourced from Brazil. Its grain can be straight or wavy, accentuating an attractive olive-brown colour.

Jarrah is another exotic, reddish-brown hardwood, resembling mahogany. It originates from Australia.

Jatobá has a coarse, interweaving grain that makes it exceptionally durable. This wood is also known as Brazilian cherry.

Santos mahogany from Brazil (as opposed to the less durable Honduran variety) is incredibly hard, with a deep reddish-brown finish.

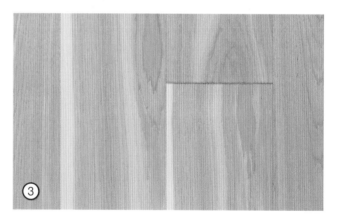

① Light hickory
② Dark hickory
③ Fawn hickory
④ Ipê
⑤ Jarrah
⑥ Jatobá
⑦ Jatobá and oak
⑧ Jatobá and wengé
⑨ Spanish mahogany
⑩ Santos mahogany
⑪ Chestnut mahogany
⑫ African mahogany

Oak is the classic wood for flooring because it looks great and performs well. Since it is one of the strongest hardwoods, it is particularly suitable for areas like kitchens, where objects can be dropped and you want to minimise the damage that can result. Oak has a prominent, grainy texture that gives it a warm character and it readily absorbs finishing materials, making it a highly versatile option.

Various types of this species are used for flooring, which vary widely in the tones they create. Red oak grows all along the east coast of North America and has warm, pinkish hues. White oak grows in most of the USA and southern Canada. It has a slightly creamy or greyish cast and is harder than red oak.

① Fine-grain oak
② Dark oak
③ Varied-grain oak
④ Pale oak
⑤ Desert oak
⑥ Oak with large knots
⑦ Oak with small knots
⑧ Oak and merbau
⑨ Wide-grain oak
⑩ Brown oak
⑪ White oak
⑫ Red oak

Maple is creamy with reddish-brown accents and a subdued grain. It is often used in schools because its hardness makes it ideal for high traffic.

Merbau is deeply shaded, very hard and considered one of the best choices for wood floors covering radiant heating systems.

Paduak (or African paduak) is reddish with shades of orange and purple. This colouring, combined with its coarse texture, makes it distinctive.

Pine contains plenty of swirls and knots and is yellowish-brown in colour.

Rosewood is full of character. Its deep brown colouring features black streaks, making it appear luxurious.

Sapeale (or sapele) is a fine-textured African wood. Its interlocked grain creates a rich finish.

Walnut is a premium-priced product with considerable colour variation and it always features a grain pattern. It matures to a lustrous colour.

① Maple
② Hard maple
③ Manchurian maple
④ Natural spalted maple
⑤ Merbau
⑥ African paduak
⑦ Pine
⑧ Rosewood
⑨ Sapeale
⑩ Walnut
⑪ Antique walnut
⑫ Butternut walnut

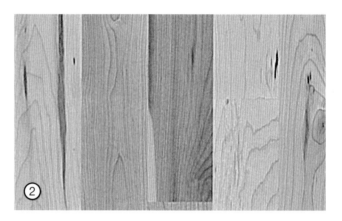

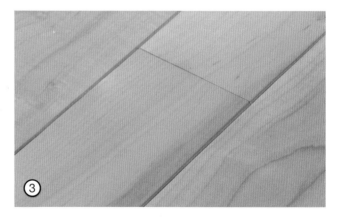

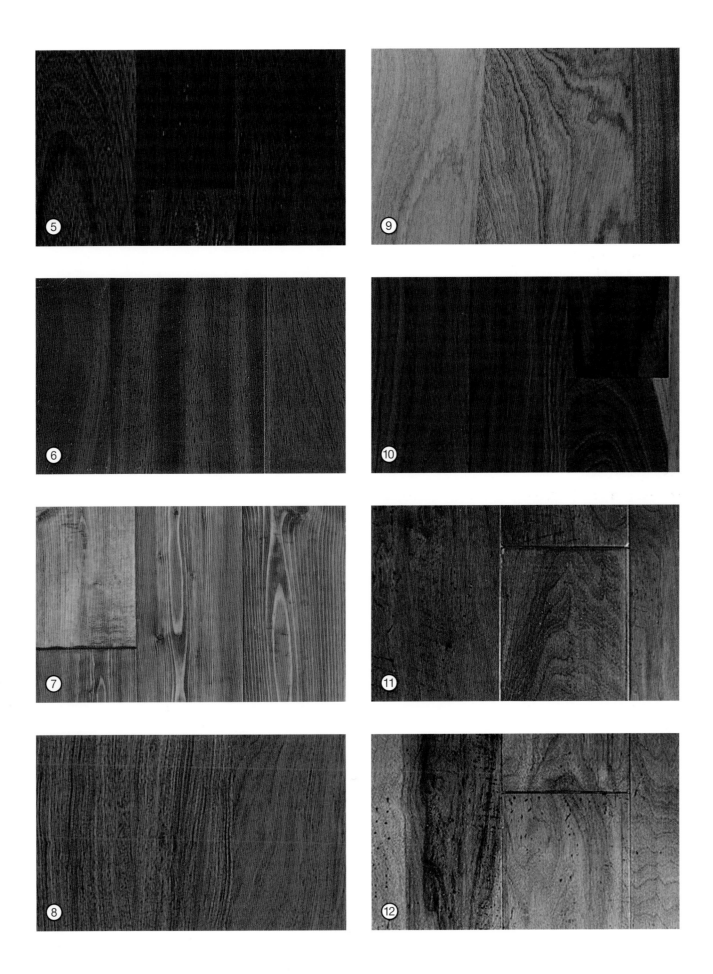

Board Choices

If the wood you choose has a distinctive grain pattern or texture, using it in wide, handcrafted planks of random widths will emphasise this characteristic. Another way to introduce character is by using two contrasting woods to give a variety of tones. Solid wood usually comes in random lengths with tongue-and-groove edges (however, reclaimed wood is usually straight-edged). Engineered wood boards, comprised of a layer of hardwood attached to a softwood base, are a less expensive alternative.

① Create contrast while still maintaining unity by staining in strips. Here, strips of the bamboo floor have been stained gold and natural to create the impression of wide boards.

② Long strips allow the beauty of deeply hued woods such as this wengé to shine through without distraction.

③ Accentuate a prominent grain with wide boards, as in this oak floor.

④ Flooring that is more than three inches wide is known as 'plank' flooring and is generally laid with staggered ends to create a pattern. The plank flooring here is butternut walnut.

⑤ Pale finishes such as birch are fairly anonymous. It is imperfections such as knots that bring character.

⑥ This merbau floor exploits natural variation in colour. Laying the floor in long strips allows the contrast to continue across the room.

Parquet

Also known as 'block' flooring or wood tiles, parquet has been used to create uniquely patterned floors for hundreds of years, with some of the finest examples found in French châteaus of the 17th and 18th centuries. Parquet is created from small strips or blocks of hardwood that are laid into repeating patterns. These patterns can be fairly simple, such as the classic basketweave or herringbone styles, or extremely complex. The many patterns can be highlighted with the use of two or more contrasting woods to create new configurations. Such is its versatility that, in design terms, parquet flooring is closer to ceramic tiles than wood. It does, however, create insistent geometric detailing that draws attention to the floor, so parquet might not suit rooms with 'busy' décor.

① A classic herringbone design in santos mahogany is shown here. This material is suitable for high-traffic areas because of its resistance to abrasions.

② Parquet flooring plays with angles and lines, adding interest without distracting from the rest of the décor and furnishings.

③ Here, a herringbone pattern is featured in an auburn-stained, red oak floor, with attractive results.

④ A large-scale version of the 'basketweave' pattern, with blocks of strips set at right angles, is shown here in red oak with an amaretto stain.

⑤ This pattern is known as 'swirl'. The middle sections have been stained to increase contrast throughout this large room.

Parquet flooring has straight lines and junction points which, viewed as small samples, can seem harsh and angular. However, in a larger-scale room this effect is changed by the character of the wood – its colours and grain patterns softening the overall effect. Parquet can seem much more intimate and welcoming than row after row of floorboards. You should view wood samples in natural light, in the room in which they will be laid, to see how they will work there.

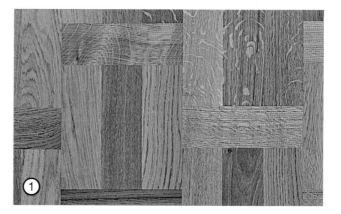

① Oak in a triple-block pattern
② White oak in a herringbone pattern
③ American oak in a triple-block pattern
④ Beech in a herringbone pattern
⑤ Jarrah in a herringbone pattern
⑥ Maple in a herringbone pattern
⑦ Rustic maple in a herringbone pattern
⑧ Merbau in a herringbone pattern
⑨ Teak in a double-block herringbone pattern
⑩ Rhodesian teak in a herringbone pattern
⑪ Walnut in a herringbone pattern
⑫ Wengé in a herringbone pattern

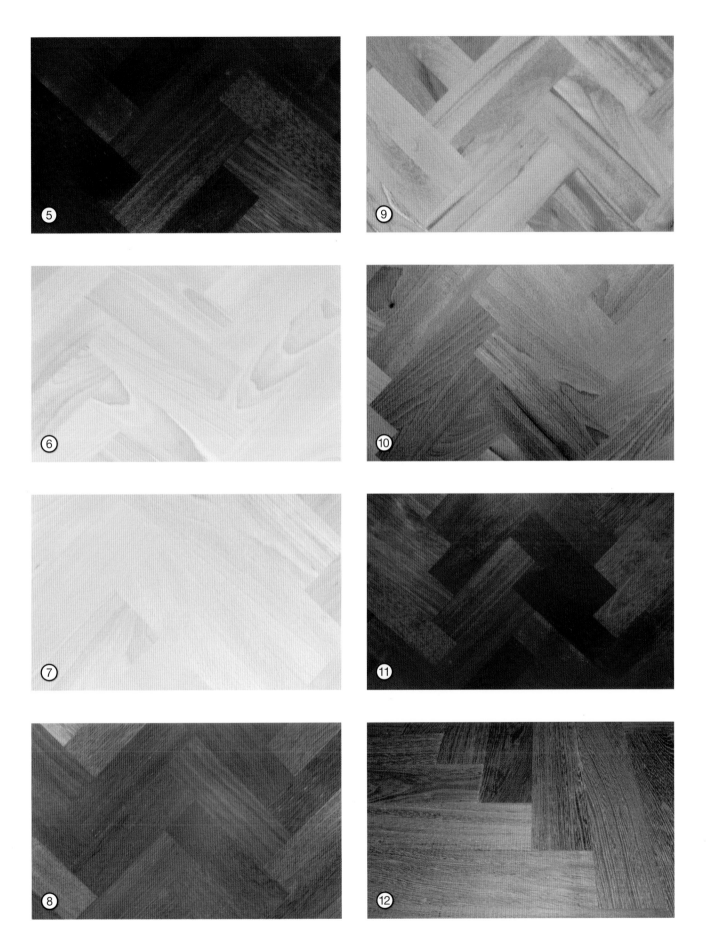

Laminate Flooring

Laminate flooring is an extremely popular choice because it can mimic materials such as wood and stone, but at less expense. Laminates are made up of layers – the top is a thin veneer, or a photographic layer that replicates the look of wood – bonded together to make a durable floor that is relatively easy to install, usually by 'clicking' sections together. Developed in Scandinavia toward the end of the 20th century, laminates have rapidly cornered the market for inexpensive, easy-maintenance flooring.

① Laminate is a great impostor. Here, a laminate floor replicates the much pricier finish of teak.

② Laminates suit most settings, including this office environment. To protect the top layer, use rubber wheels or protective mats.

③ A laminate in a natural varnished oak plank won't break the budget, and no one will know the difference.

④ Maple is a popular choice for a light wood effect, which is achieved here by a laminate that beautifully unites the wicker storage baskets and the neutral décor.

⑤ Vintage aged elm looks great but costs a fortune. A copycat laminate can look just as good at a fraction of the price.

④

⑤

Laminates allow you to have a 'wood-looking' floor at a fraction of the price of the real material. The introduction of a new breed of laminate that has a thin veneer of real wood as its top surface guarantees that the floor will look realistic, because it is. However, if you opt for a laminate with a photographic layer, look for flooring that uses embossed in register (EIR) technology, with a superior impression of real hardwood and improved slip-resistance. Because there are no joints, laminate does not trap dirt. However, dirt and other materials may still collect and be ground into the top layer, so regular sweeping and occasional mopping are required. Laminate is still easier to maintain than wood, which does not cope well with water. In some ways, laminate is even better than the real thing!

① Vintage oak with dark varnish
② Vintage oak with natural varnish
③ Oak in colonial planks
④ Oak planks with natural varnish
⑤ Oak planks with dark varnish
⑥ Wild maple in colonial planks
⑦ Wild maple in amber planks
⑧ Wild maple with natural varnish
⑨ Merbau
⑩ Teak planks with natural varnish
⑪ Oiled walnut planks
⑫ Wengé

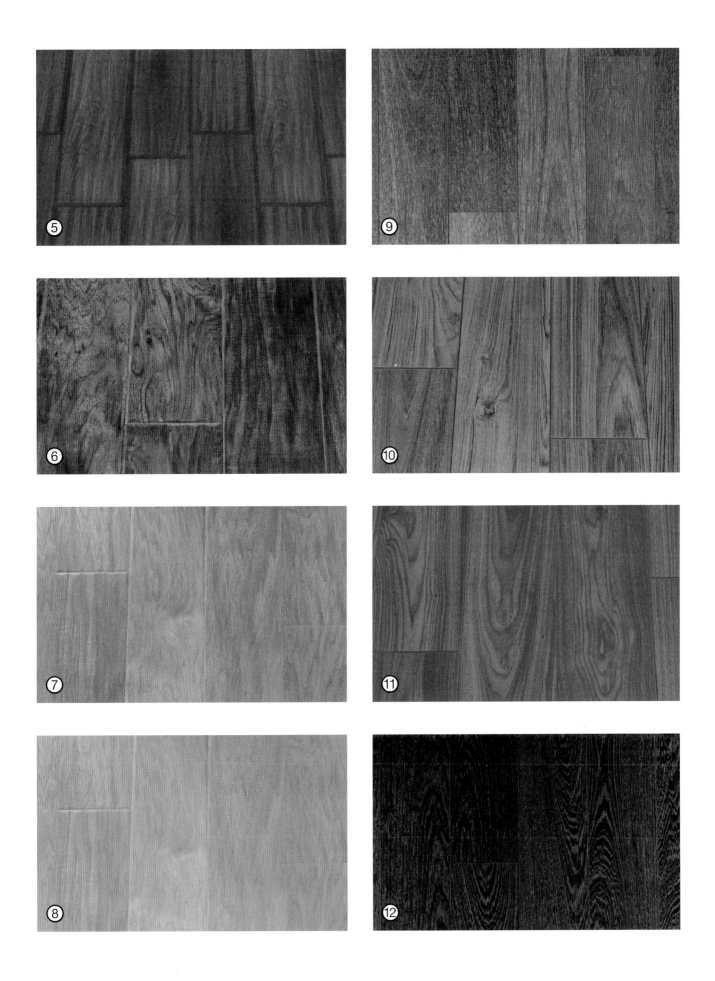

Decorating Floors

With their inherent beauty and the patterns and textures in their grains and knots, wooden floors are very attractive. There are times, however, when the effect can be improved by adding visual interest. The main options are staining and stencilling. Stains come in a variety of colours or can simply be used to darken the wood's tone without concealing the grain. The easiest to apply are still the oil-based variety, which also tend to produce better results. Stencilling allows you to add a motif, perhaps as a border or to create a focal point.

Drawing patterns free-hand is another decorative option. This option is especially appealing if the floor is showing signs of age, since you can paint over the marks of wear and tear. Use the highest quality deck paint, because of its durability, and draw patterns by hand, or use the board widths as your guide and create a draughtboard design. Another option is spattering, in which a pebblelike finish is created by painting the floor and flicking mineral spirits over it before the paint dries.

① Staining and stencilling are combined to create a unique green floor with repeated diamond motifs.

② Bleaching lightens the tone of wood and creates a distressed effect. This is useful if you want to change the look of a wood floor that has darkened with age.

③ Painting floorboards disguises any flaws and allows you to create a completely new look – but signs of wear can show up quite fast.

④ Painting a draughtboard design produces a new pattern while retaining some of the original character of the wood.

Staining can completely change the character of the wood, as these pictures illustrate.

⑤ Natural bamboo
⑥ Walnut-stained bamboo
⑦ Cherry-stained bamboo
⑧ Charcoal-stained bamboo

① A stencil is a cut-out from card or acetate and can be as simple or complex as you wish.

② Use two colours to create contrast and make the border more attention-grabbing.

③ An (almost) continuous line leads the eye around the room, making it seem bigger, while curves soften the otherwise harsh effect.

④ Use an accent colour such as this red to pull the eye towards the corners of the room, giving an impression of depth.

⑤ When put together, simple motifs will create an intricate design.

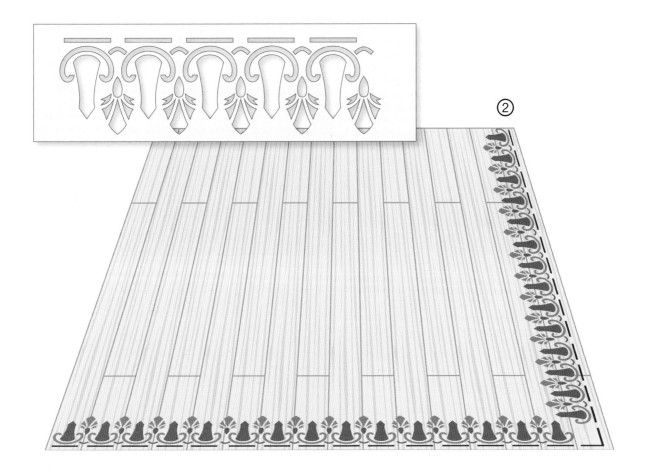

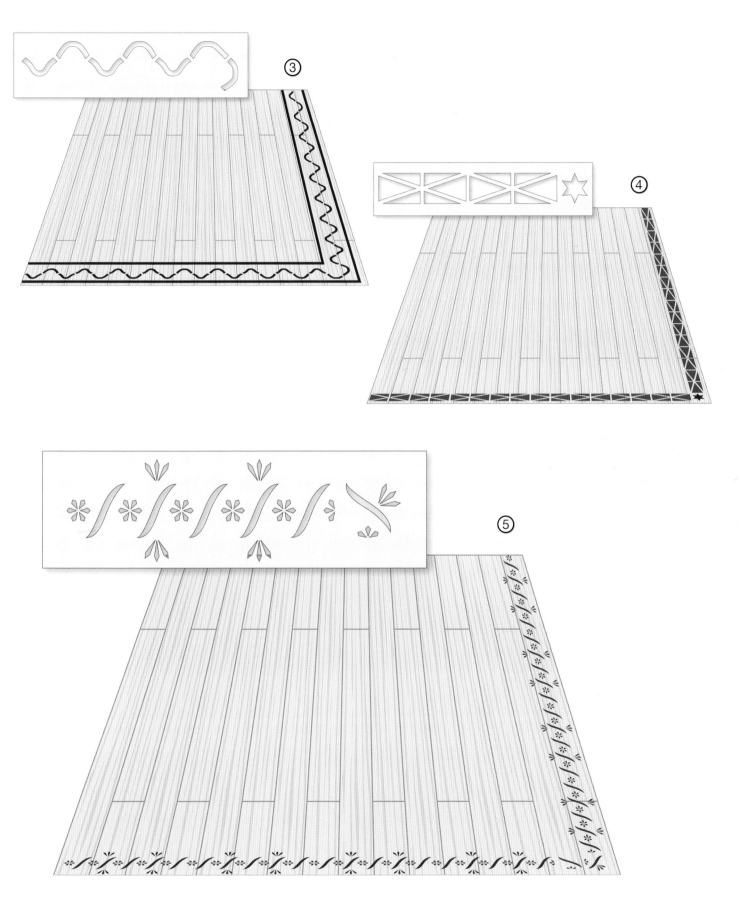

③

④

⑤

Using Wood to Make Patterns

There are countless parquet patterns and very little consistency in what they are called; in fact, the same name may be used to describe two very different geometric designs. For this reason it is crucial to see a sample of what you are buying from your chosen supplier. The most common patterns are available in oak, teak, rosewood and various other exotic species, but you may be able to put in a special order for your chosen combination – although this may cost more.

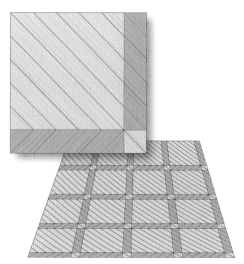

Diagonal lines

Herringbone variation

Classic herringbone

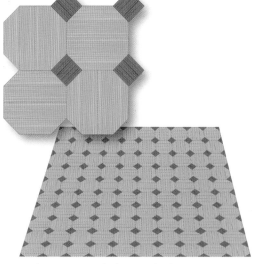

Basketweave with diamond inlay

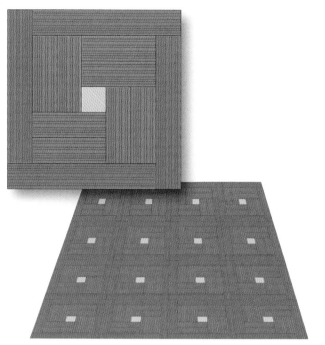

Swirl with inlay

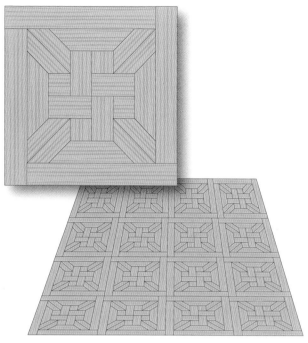

Geometric cross

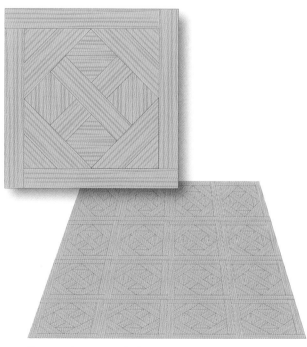

Geometric cross variation

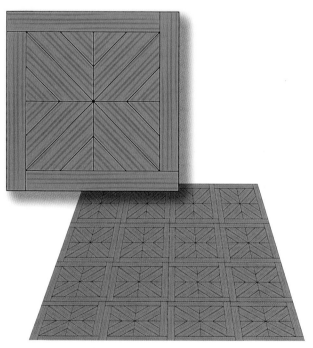

Converging diagonals

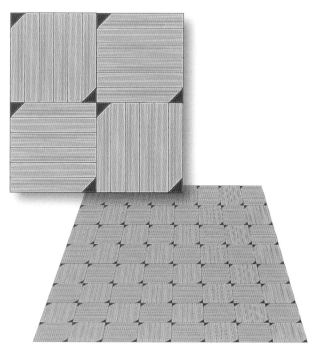

Basketweave with corner inlays

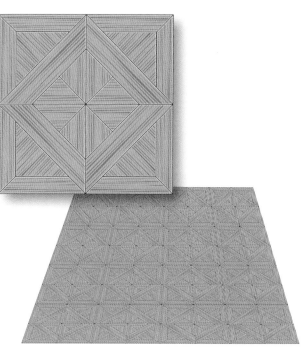

Gothic

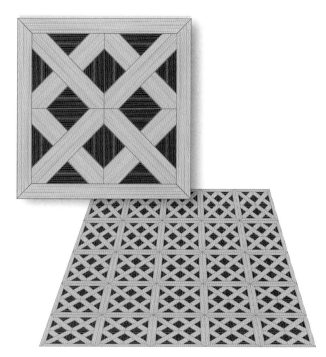

Lattice effect

Tumbling blocks

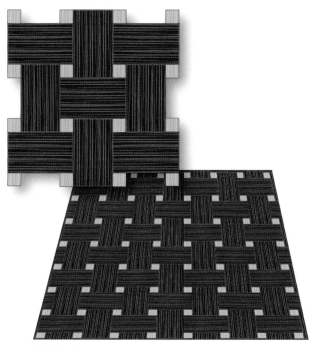

Half basketweave

Basketweave with border

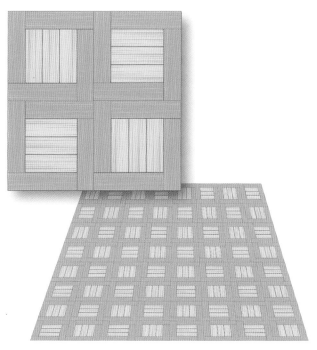

Basketweave with borders

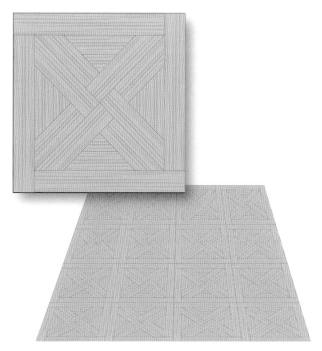

Interwoven cross

Using Colour

Colour sets the mood for a room and adding colour to a wooden floor will have a major influence on the rest of the décor. Left untreated, the finish is likely to be a natural light or golden brown that will darken with age. If the finish is relatively uniform, stripes and textures will add visual interest. If you prefer a harmonious scheme, use nature – the colours of the sea and sand, bark and leaves, or stone and moss – as your inspiration. For more contrast, consider the colours of exotic blooms or rich spices. When planning a scheme with a wood floor, retaining the pale, neutral tones can emphasise the material's natural origins, as well as soften the hard lines in a room. In contrast, as in the photographs opposite, dark wood can add depth and drama.

① Pale blue is a cool colour that balances the warmth of the wood, creating an interesting contrast.

② White harmonises with any colour, creating a light, airy effect.

③ Lemon walls will bring out the yellow tones in the floor, and the deep blue cushions prevent the room from feeling overly cute.

④ The leafy look of pale green walls instils harmony and blends in with a wood finish to unify the room.

⑤ Burnt-orange walls draw out the drama of this dark wood. The look is balanced by a neutral, white wall.

⑥ Bold purple furniture emphasises the warmth of this floor. The rest of the room is painted white to avoid clashes, resulting in the calmest overall effect of all of the choices shown here.

⑦ Pink walls contrast with the deep brown of the floor, shaping the lines of the room and adding a new, lively feel.

⑧ In this setting, strong citrus yellow and green bring a sense of freshness and vitality.

Resilient Floors

A resilient floor is the Cinderella of flooring: hard-working, but largely ignored or treated with disdain. It is easy to install and maintain and it offers comfort and value. Although vinyl is the most common choice, fellow resilients linoleum, rubber and cork are gaining ground because of their environmental benefits. As well as being made with natural materials, linoleum has the added advantage of being nonallergenic.

Opposite: Linoleum can be custom-made with borders and inlaid patterns to meet the requirements of an individual setting as here, in this elegant living room.

Below: Whatever colour or pattern you choose, chances are you can find it in vinyl form and vinyl can be used in all types of rooms.

Resilient means 'springs back into shape readily'. This quality is one of the most useful characteristics of this type of flooring. It is also comfortable to walk on and any items that are dropped have a chance of surviving their fall. Resilient floors come in colours, designs and effects to suit every budget and setting. Vinyl and linoleum are especially good at mimicking the appearance of other materials. They won't sound, smell, or feel like the real thing, but they won't cost as much either. Being softer, most resilient floors will wear out: vinyl doesn't last more than ten years, though linoleum survives longer.

Because it is easy to clean, cheap and scuff-resistant, resilient flooring is mainly used for kitchens, laundry rooms, utility rooms, bathrooms and playrooms. It is worth investing more if the application is in a high-traffic area such as the kitchen. Thicknesses vary; the deeper the material the better it will perform.

Vinyl is the most popular resilient floor because it can be supplied in a wide range of colours and textures and it is coated with a protective layer that renders it virtually maintenance-free. If you want striking designs or a classic look without the huge budget, this is a sensible choice. It loses some popularity because it is not environmentally friendly at all, being made from plastic. Linoleum, cork and rubber, on the other hand, are all created from natural products or are simply natural, sustainable materials in themselves. A luxury addition to this sector is leather, a resilient material that can create a stunning floor.

Resilient flooring can be applied over a variety of surfaces, including plywood subfloors, concrete, wood and linoleum. It is important that the surface is flat and sound, as any imperfections might be felt underfoot and can damage the soft material – most resilient materials are susceptible to scratches, dents and burns. If you take this consideration into account, your 'green' floor will give you years of faithful service!

Types, Sizes and Finishes

Most types of resilient flooring are available as either sheets or tiles. Sheets are much easier to install, though precise cutting is required for edges and obstacles. Tiles are more flexible and good for dealing with complicated spaces, but they require a good adhesive and careful planning to prevent gaps from appearing between them. Tiles allow you to mix and match colours, creating your own unique effect. What the finished floor will look like comes down to what you desire – in appearance at least, if not in texture.

① Cork can be stained to make borders or define particular areas, as in this bedroom.

② Wood-effect vinyl is available in random-sized planks, to mimic the real thing even more realistically.

③ For a stone floor on a budget, look no further than vinyl.

④ The soft colours and design flexibility of linoleum suit it to any room, but it works especially well in bathrooms.

⑤ Some types of linoleum can be coloured, cut and laid any way you want; here, to create inserts.

⑥ Linoleum shapes appear again here. The result is a truly individual floor.

Cork

Cork is one of the most environmentally friendly flooring materials, as it is a natural product taken from the forests of Portugal and Spain. The bark of the cork oak tree can be cut away every nine years without causing damage, which means that it can be sustainably harvested. Forget about the soft, fragmented cork of notice boards: cork is rugged and incredibly durable. Because about half of its volume is air, it feels soft underfoot and has superb acoustic and thermal insulation properties. It also contains suberin, a substance that protects it from moisture damage, as well as being a natural insect repellent. Cork is also hypoallergenic, making it an excellent material in the homes of allergy sufferers.

① Cork can be laid in small sections to replicate the look of parquet flooring.

② Mixing some of the different colours available creates a lively and inviting room.

③ There is a surprising range of colours available in cork, including this elegant grey.

④ Cork can be stained to create inlaid effects, defining parts of the room.

⑤ An inlaid flooring pattern becomes the focal point in this otherwise neutral kitchen.

③

④

⑤

Cork comes in a range of earthy finishes, including a wealth of natural tones (which show off the grain) and a range of stained hues, such as charcoal, green, blue and even grey. It is sold in thin tiles or thicker, interlocking planks, so you have the option of mixing different colours to create a draughtboard pattern. A polyurethane coating is required every few years to avoid stain and water damage, as cork cannot be sanded.

① Natural
② Mottled honey
③ Harvest blocks
④ Mottled gold
⑤ Mottled natural
⑥ Flecked coffee
⑦ Fawn strips
⑧ Flecked gold
⑨ Dark wood strips
⑩ Red wood strips
⑪ Mottled green
⑫ Black blocks
⑬ Flecked cherry
⑭ Light wood strips
⑮ Flecked sage
⑯ Grey blocks
⑰ Flecked light wood
⑱ Mottled light wood
⑲ Gold blocks
⑳ Mottled dark wood

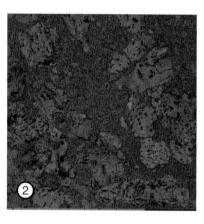

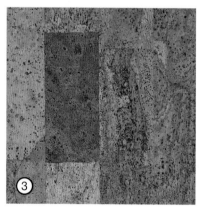

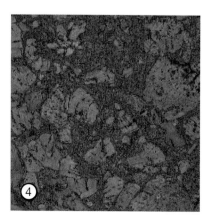

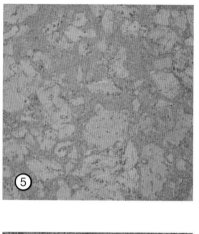

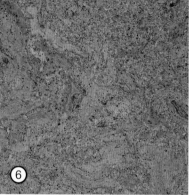

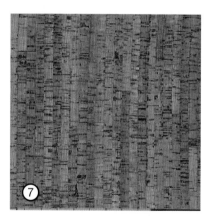

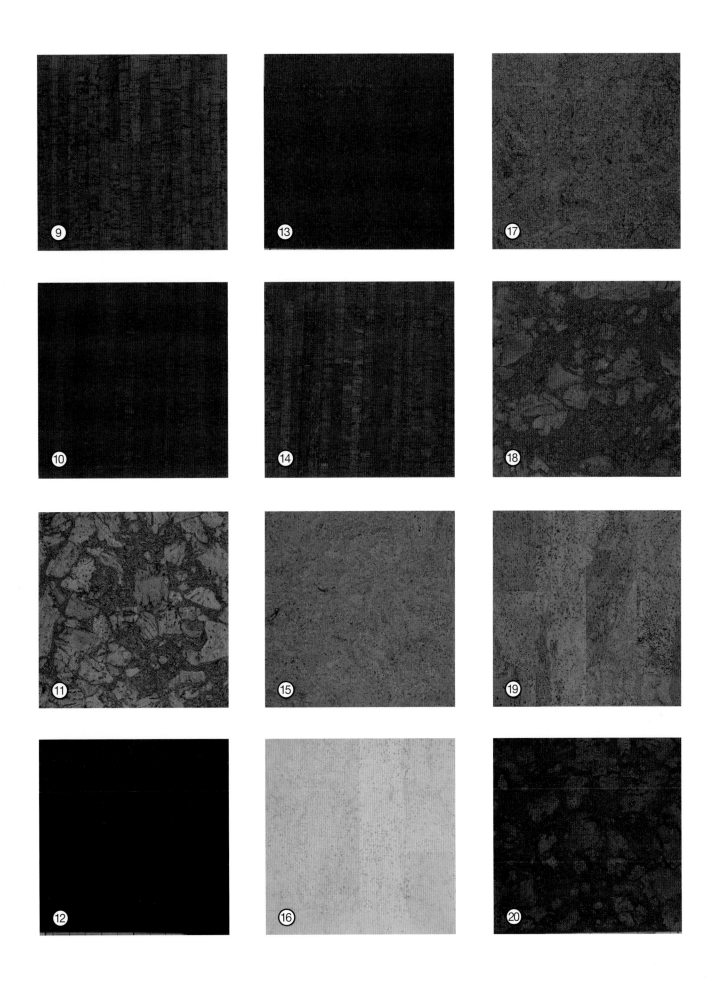

Leather

If you want truly luxurious, unusual flooring, and your budget will allow, go for leather. It brings a sense of sophistication, warmth and softness – and, of course, it is a completely natural product. A leather floor will mature gracefully with age: its rich colours will deepen and it will accumulate a patina of blemishes that will add a sense of history and longevity to the room. It is totally unsuitable for kitchens and bathrooms (leather and water don't mix) but marvellous for the living room, study or library that needs a touch of class.

① Light shades of leather closely resemble wood (with a richer lustre), so the two materials pair well in settings such as this bedroom.

② Leather flooring complements the colours and textures of leather furniture, giving this room a natural warmth and sophistication.

③ Leather can be dyed almost any colour, allowing for the creation of this unusual green-and-black draughtboard tile floor, laid at an angle so that it flows across the room.

④ Leather and wood are matched in this living room with its flowing, curved furnishings.

⑤ This stairway gets the 'red carpet' treatment with a lavish leather runner. The border pattern echoes the twists and curls of the wrought-iron balustrade.

Leather flooring is supplied in tiles, which are usually square but can be rectangular, hexagonal, or octagonal if you choose, since they are hand-cut. The finish can be smooth or antique-textured (like the faux animal hides on the opposite page, for example) and will need to be waxed and buffed regularly. Colours vary within batches and range from beige to black, including all shades of brown and some shades of green and red.

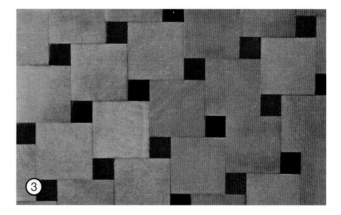

① Victorian etched border
② Red and gold diamonds
③ Contrasting red and black squares
④ Geometric gold brick
⑤ Tan alligator skin
⑥ Brown calf skin
⑦ Black hippo skin
⑧ Khaki lizard skin
⑨ Brown beaver skin
⑩ Green elephant skin
⑪ Burgundy hardboard
⑫ Floral fawn

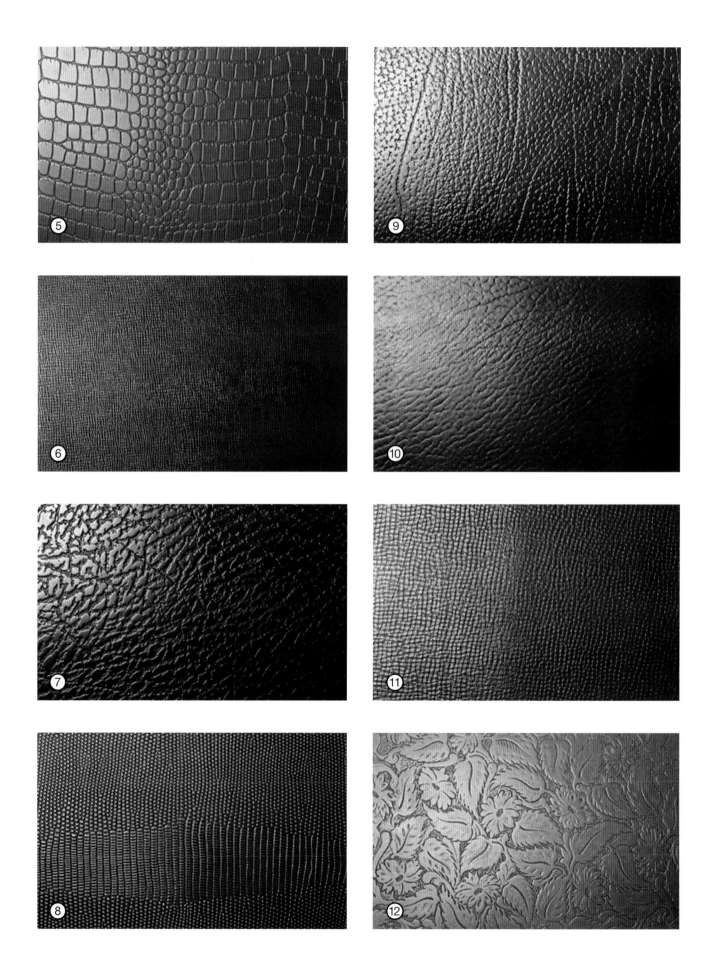

Linoleum

Linoleum is frequently confused with vinyl flooring but it is actually very different, as it is much more durable and made from natural ingredients. Linoleum is a solidified mixture of linseed oil, wood and cork powder, pine resin and ground limestone that is spread on a jute or burlap backing. Popular in the late Victorian period as an inexpensive, hard-wearing floor, it fell out of favour because it cracked easily, and vinyl took its place. Its enormous range of colours and improved quality, combined with its 'green' character and antistatic, antibacterial benefits, have brought linoleum back with a vengeance.

① Blend a range of colours to create texture using linoleum tiles, as in this contemporary living room.

② Linoleum actually gets stronger as it ages, so it is perfect for high-traffic environments, like this entrance hall.

③ Linoleum is excellent for bathrooms, as long as water cannot penetrate it: abut tiles tightly or hot-seam the sheet variety.

④ Linoleum has a natural, soft character that makes it ideal for use in kitchens.

⑤ Invented in Victorian times, linoleum can fit in with period furniture in a way that vinyl cannot.

Linoleum has a naturally grainy, matte finish that can be pigmented to create an enormous range of colours and effects, such as marble, streaks and flecks. These pages show plain colours in the popular marbled finish.

① Forest
② Grey
③ Ivory
④ Scarlet
⑤ Deep blue
⑥ Charcoal
⑦ Pale blue
⑧ Grey and blue
⑨ Mottled blue
⑩ Coffee
⑪ Ochre
⑫ Gold

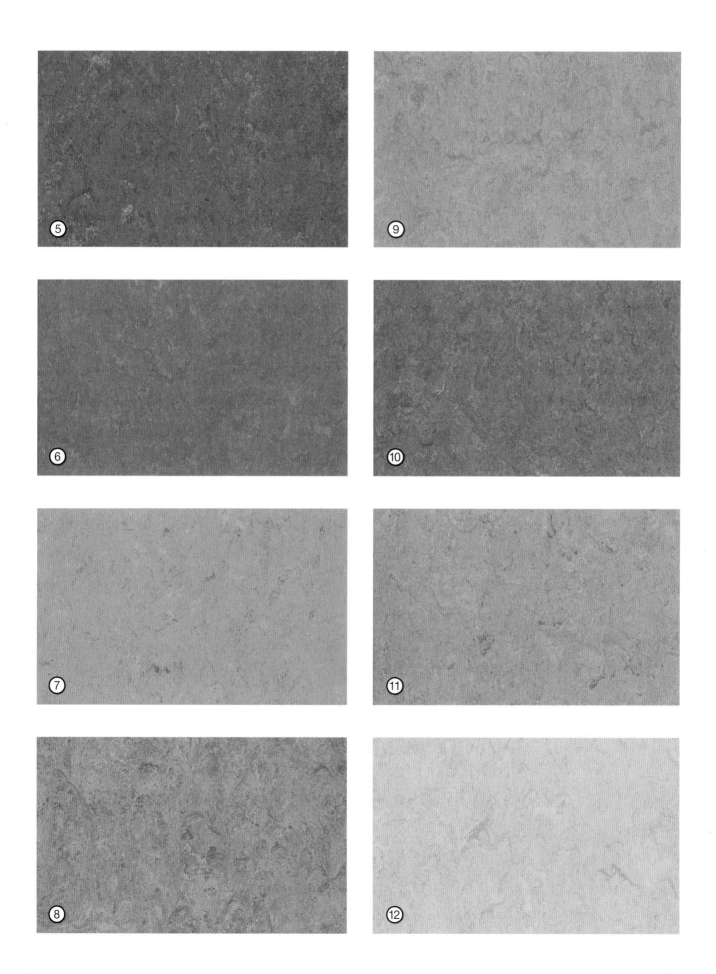

Continued from page 180.

Linoleum can be produced in a vast range of patterns (geometric effects, for example) and with a choice of borders. Precision cutting using computer technology now allows the creation of complicated inlaid designs and motifs, such as feature panels and inset squares.

① Geometric pattern
② Simple border
③ Faux tiles in 'copper'
④ Faux marble
⑤ Stone-effect inlay
⑥ Faux granite
⑦ Inset squares
⑧ Spattered effect
⑨ Aged brown
⑩ Faux oak
⑪ Burnished blue
⑫ Faux metal insets
⑬ Faux stone parquet
⑭ Faux walnut
⑮ Crazy paving
⑯ Faux mosaic tiles in 'gold'
⑰ Faux mosaic tiles in 'moss'
⑱ Faux beech
⑲ Faux glass
⑳ Faux mosaic tiles in 'chocolate'

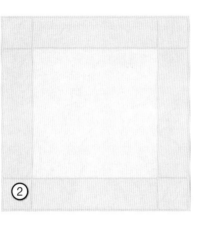

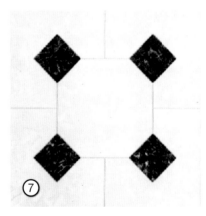

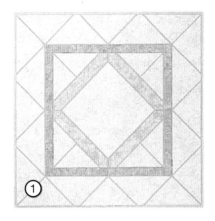

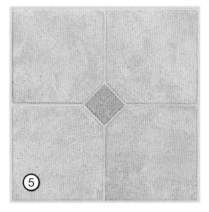

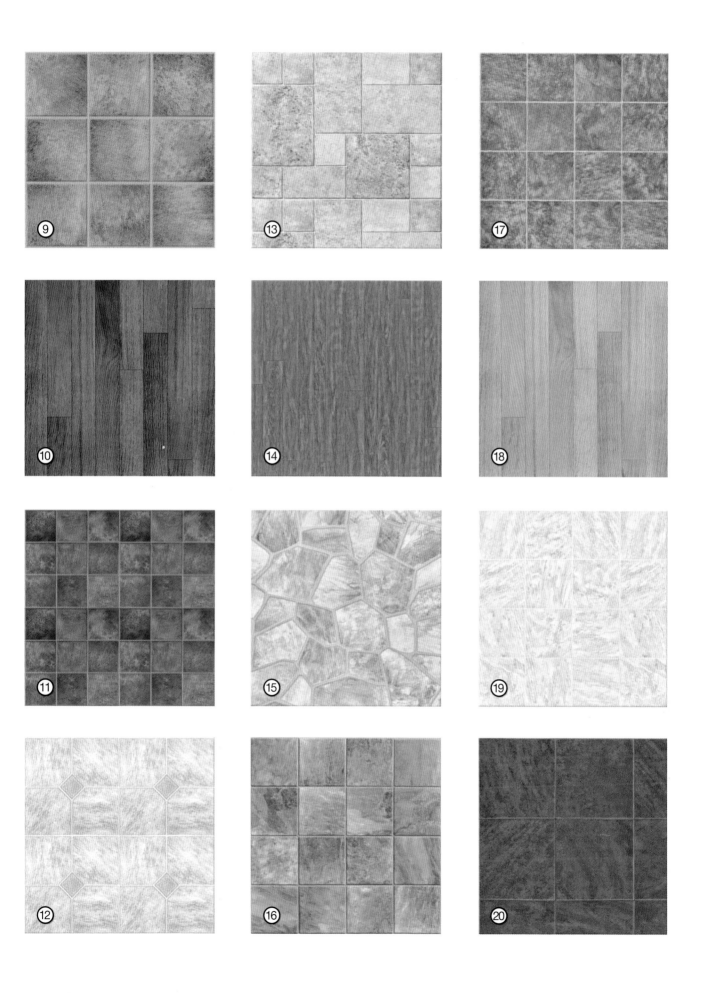

Rubber

For durability and high performance in a range of colours, rubber is a superb choice. Sometimes identified with industrial and commercial applications, this material is also fantastic for many domestic settings because it is so versatile and hard-wearing (a rubber floor should last 20 years). Although virgin rubber can be naturally harvested, most rubber flooring is produced synthetically and there is increasing availability of this material in recycled form. The synthetic kind (made from vulcanised synthetic rubber, pigment, silica and china clay) is much better than natural rubber at resisting fats and chemicals – a crucial point when it is to be laid as a kitchen floor. Rubber is quiet underfoot and easy to maintain, and its tendency to become slippery when wet can be counteracted by choosing textured or studded finishes.

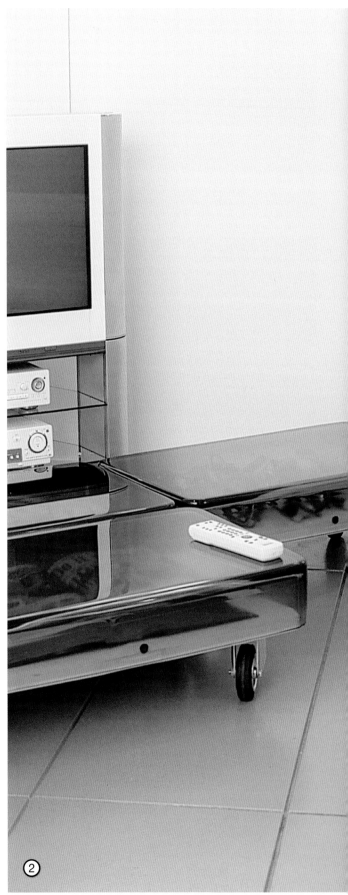

①

②

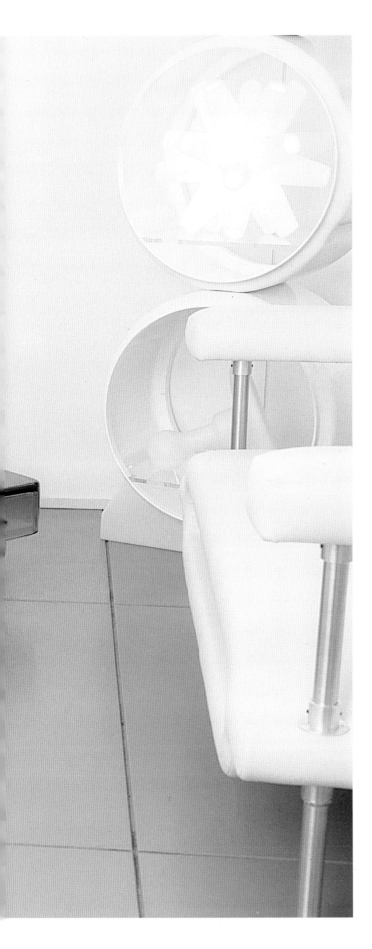

① A high-sheen, polished finish suits industrial-style settings, like this contemporary kitchen.

② In this living room, a matte finish softens the occasionally harsh, sterile character of the material.

③ Rubber is also a practical and stylish material for bathroom flooring, though bright colours like this striking magenta might not be stocked by all suppliers.

Rubber flooring is available in sheet and tile form. It can be polished to a high sheen or left matte for a warmer feel. The surface can be raised to create dimples, studs and other patterns for more texture and greater slip-resistance. No other flooring material combines softness and durability – qualities that have made rubber perennially popular in commercial applications – in quite the same way. Some suppliers stock a range of colours, which provide variation from the bland greys usually chosen for such industrial settings.

① Black studs
② Blue studs
③ Purple studs
④ Lilac studs
⑤ Sage mica
⑥ Emerald mica
⑦ Dark coral hammered
⑧ Turquoise hammered
⑨ Lime raised studs
⑩ Magenta raised studs
⑪ Gold dot smooth
⑫ Orange zig-zag

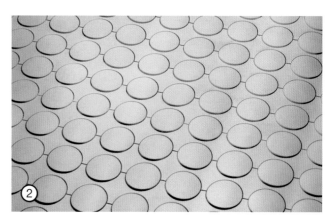

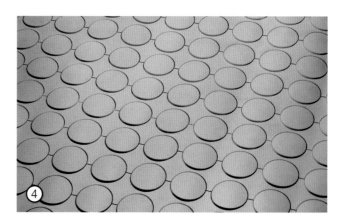

Vinyl

Vinyl is a practical and popular flooring material because it comes in such a wide array of colours and patterns, it performs well and it is easy to maintain. Made of plastic, vinyl isn't environmentally friendly, but it is very kind to your budget. In fact, vinyl mimics other more expensive materials quite successfully. Its hard-wearing and slip-resistant characteristics combined with comfort underfoot have seen it laid in countless kitchens, bathrooms and utility rooms.

① Laying a mosaic floor in this large bathroom would be expensive and time-consuming, and chances are the subfloor couldn't take the weight. Vinyl comes to the rescue, replicating a tile finish admirably.

② It won't feel like wood, but it does look like it: vinyl can mimic almost any material and is a great choice for those on a tight budget.

③ A wooden slat design brings warmth and a cosy feel to this kitchen and living area.

④ Vinyl replicates the look of stone very well, though not the texture. The giveaway is often the two-dimensional effect, which exposes the lack of genuine grout.

⑤ Vinyl offers countless decorative finishes, such as this ivy effect, allowing you to find an unusual look that suits your home and décor.

⑥ Vinyl is an easy-to-maintain option for rooms that will see a lot of wear and tear, like this games room.

Vinyl's key selling point, aside from low price and ease of maintenance, is the variety of colours and finishes that are available. It can resemble many kinds of wood or stone and often features insets or borders. You can opt for innumerable colours and effects. Vinyl comes in tile and sheet form in a variety of grades, depending on thickness. The thicker grades of vinyl sheet have a deeper PVC cushion layer plus a urethane top layer. This layering makes the floor softer underfoot and more successfully covers flaws in the subfloor.

① Faux granite
② Faux terrazzo, dark
③ Faux slate
④ Faux rough plaster
⑤ Faux terrazzo, light
⑥ Faux limestone
⑦ Faux pink marble
⑧ Mottled finish
⑨ Draughtboard pattern
⑩ Intricate inlays
⑪ Geometric pattern
⑫ Faux beige tile
⑬ Faux white marble
⑭ Faux sandstone
⑮ Natural stencil pattern
⑯ Faux red sandstone, with inset
⑰ Sky effect
⑱ Faux leather
⑲ Faux tiled insets
⑳ Faux mosaic tile

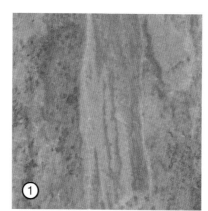

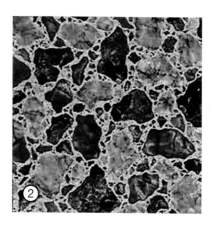

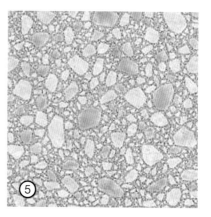

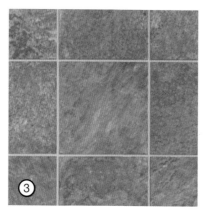

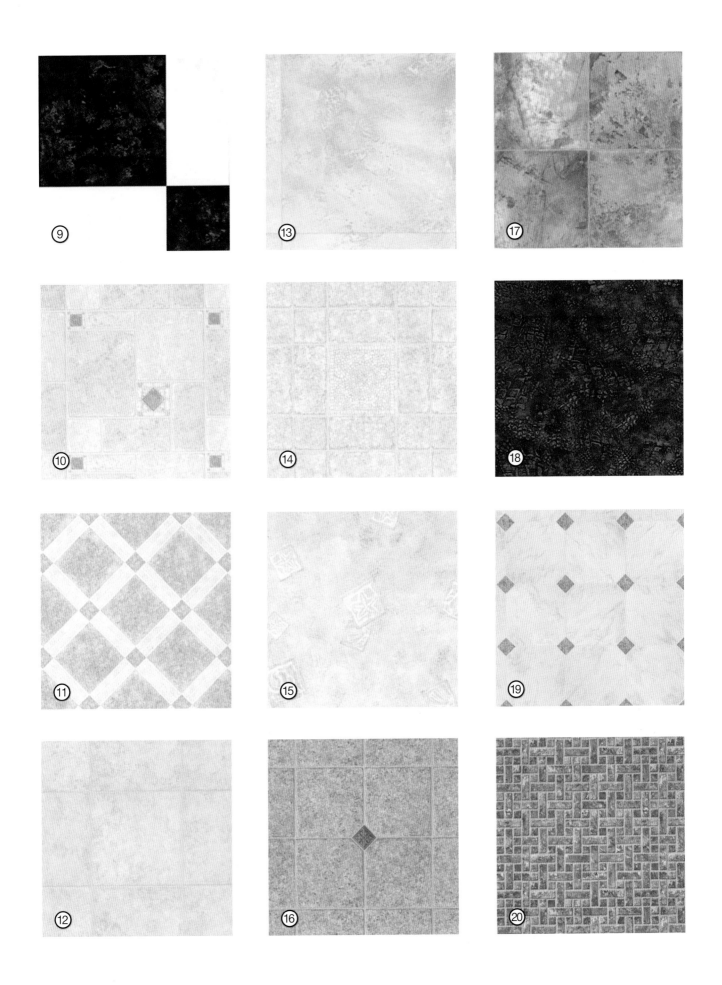

Continued from page 191.

① Faux exotic wood
② Faux birch
③ Faux glass mosaic
④ Neutral leaf pattern
⑤ Faux mahogany
⑥ Faux oak
⑦ Faux riven slate
⑧ Autumnal leaf pattern
⑨ Faux bamboo shoots
⑩ Faux blue slate
⑪ Dot pattern
⑫ Faux terracotta
⑬ Faux ceramic tile in grey
⑭ Faux cork
⑮ Faux black slate
⑯ Faux white slate
⑰ Faux ceramic tile in beige
⑱ Faux wooden inlay
⑲ Mottled blue
⑳ Faux brick

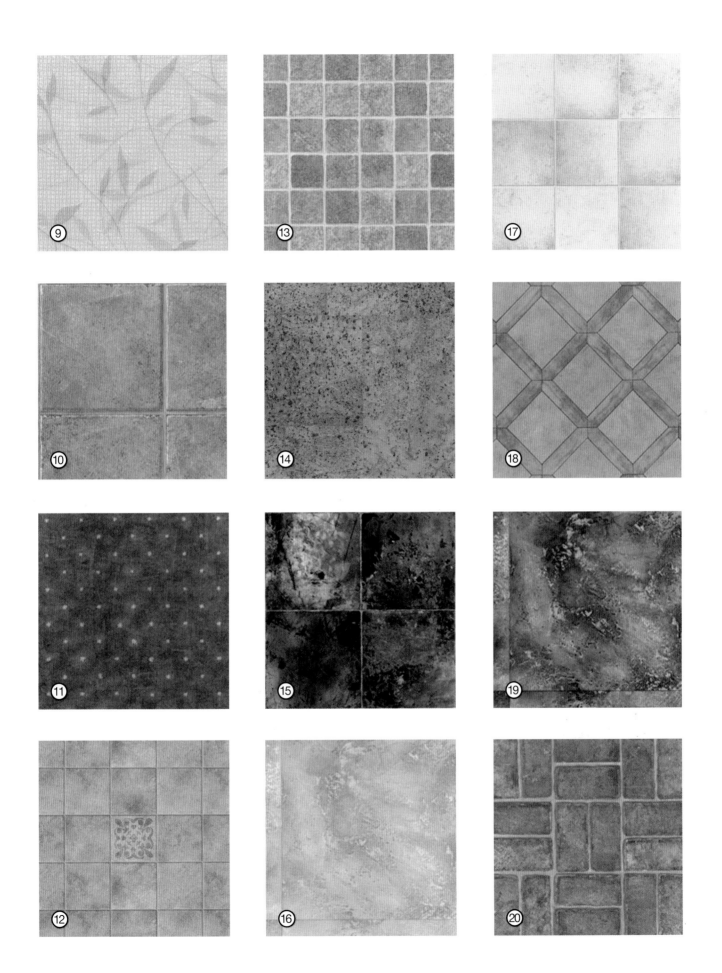

Using Colour

The range of colour in resilient flooring is so wide that it can be very difficult to reach a final decision. Consider using stripes to emphasise the length of a room, especially in kitchens. Bigger rooms will need wider stripes. Balance plain walls and furnishings with some textural interest from a patterned floor. Always bear in mind that the floor is part of an overall scheme for the room and that, while you can recolour the walls every few years, the floor will stay as it is.

①

②

③

④

① The soft blue furniture and white wall harmonise with a strong but warm, blue-and-cream striped floor.

② For more contrast with the same floor, try a pastel pink wall to add more warmth and interest.

③ This room features a dominant black-and-white striped floor. To create a sense of balance, the room also incorporates a strong retro accent in red and a complementary peach wall.

④ Let the floor surface really stand out by giving it a complementary neutral backdrop.

⑤ A stone-effect floor is paired with black and neutrals here for a contemporary look.

⑥ This is a harmonious scheme where two shades of blue – offering more depth than one shade alone – are boosted with a lilac accent.

⑦ Alternatively, draw the eye in with strong contrasts, here achieved with lemon walls and green accents against a sand-hued floor.

⑧ Take your room back to the 1950s. Here, turquoise and terracotta accents take centre stage because of the neutral floor.

Soft Floors

Carpet is one of the most popular flooring choices. It is soft and warm underfoot and offers beautiful colours and patterns, creating a focal point in a room. Carpet brings a soothing intimacy to homes, and no other material matches the luxurious feel of cushioned wool under bare feet. If you want to add character to a floor, just lay a rug down. If you like your décor to reflect simple, earthy beauty, investigate the world of natural flooring.

Opposite: Carpets and rugs add textures like no other flooring material. This quality is especially valuable when adding interest to a neutral colour scheme.

Below: Wood accompanies natural-fibre sisal carpet in this living area. The colours on the walls are bright and fresh without being overwhelming.

Carpet is a versatile flooring material that, in addition to its comfort and appearance, offers slip-resistance and insulation. This type of flooring can't cope with moisture, so it is most suitable for living rooms, hallways, landings, stairways and bedrooms.

Carpet is made in both natural fibres, such as wool, silk, jute, sisal, coir, flax and seagrass, and man-made materials – which are better at retaining colour and resisting soil and stains – such as nylon, polyester, polypropylene (trade name olefin), acrylic and viscose. The fibre mix might be 100 per cent wool (as the first 18th-century carpets were) or totally synthetic, or a blend of the two. Higher wool content brings softness, a fresh, buoyant look, good colour retention and flame-resistance – and a higher price. Manmade fibres are harder wearing, so most carpets are now predominantly synthetic. With carpet more than any other flooring material, you get what you pay for.

Carpets are graded on their durability, from heavy domestic (used in hallways and living rooms), to general domestic (bedrooms) or light domestic (spare rooms). There is also a large choice of finishes, achieved through combinations of pile choices, fibre types and the way the carpet is manufactured. All carpets were once woven, but now the majority are tufted in yarn loops to create the pile. The density of this pile dictates how the carpet will resist dirt and crushing: the higher the density, the longer the carpet will last.

The choice of colour and pattern is also important. Remember that lighter shades and patterns will show dirt and stains more, but will also open up the room to make it seem more spacious. Darker and more complicated patterns may make the room appear smaller, but they also show less dirt. Once you have made your choice, you'll have considerable comfort to look forward to!

① A very busy pattern can work in ornate settings, as long as the style is fairly regular and the colours match those of the walls and furnishings.

② Natural materials are especially suitable for areas that link to the exterior, emphasising the connections between the indoors and outdoors.

③ Rugs are available in an infinite variety of colours and designs. The subtle hues and patterns of this rug bring vivacity to a cool room.

Types, Sizes and Finishes

When shopping for carpet, be sure to check the widths. This is especially important if you want to design the floor without seams. Another crucial factor is the quality of the pad on which the carpet will be laid: this can have a huge impact on how the carpet feels and lasts. Go for the best pad that you can afford. People won't see it, but they will sense its quality! If you do not want to commit to carpet but desire wall-to-wall coverage, carpet tiles may be the best option for you.

④ Utterly impractical anywhere else, this thick pile carpet turns a bedroom into a luxurious refuge.

⑤ Use rugs to define areas, as in this cosy 'room within a room', which is part of a large living space.

⑥ Large rugs are as good as carpet in rooms with attractive floor surfaces and they are easier to lay than carpet. The lack of formal installation allows you to change the floor as often as your fashion sense dictates.

Carpets

Carpet choices are pretty much endless in terms of fibre type, colour and pattern. In fact, you'll be absolutely spoiled for choice. Fibre types other than wool are covered on page 204, while this section shows some of the colours and patterns available. Most residential carpet needs to be stretched to fit, which is a job for a professional. The durability and performance of any carpet is heavily influenced by the quality of the padding underneath. This doesn't need to be the thickest available: the issue is the combination of firmness, support and cushioning. A carpet dealer can help you choose the option that best suits your top layer.

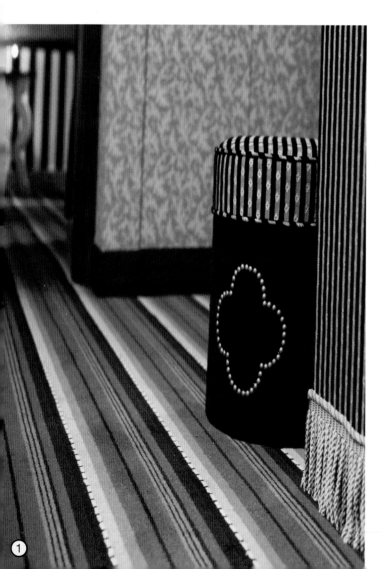

① Stripes elongate a room, especially in these bold colours, so make sure they run in the most favourable way!

② The bolder the carpet pattern, the plainer the walls can be. Repeat colours round the room, as shown in the carpet, curtains and bedspread here.

③ If in doubt, select a neutral colour that won't clash with the décor. Most neutral colours hide dirt well, too.

④ This scene is vibrant but balanced: the orange of the sofa is repeated in the carpet's abstract pattern, bringing the room together.

⑤ Plain doesn't need to mean uninteresting. The woven texture of this carpet attracts attention, complementing the detailing of the furniture.

⑥ Hallways require practical choices. As high-traffic areas they need a durable carpet that won't show too much wear. Lay a carpet with a repeated motif along the length of a hallway if you want to make it appear longer.

⑦ Carpet offers so many choices: this leopardskin design creates an exciting, exotic setting.

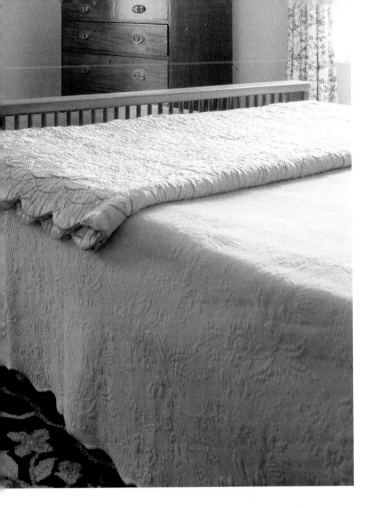

Natural wool is the most luxurious fibre because it is soft yet durable. It is also an ecologically sound option because, of course, it comes from a sustainable source – sheep! However, wool is much more expensive than synthetic materials, and the vast majority of the wool that is used in carpets is combined with artificial elements to reduce the cost and help it last longer. There are still some pure woven-wool carpets available, but make sure that you ask the dealer for these specifically.

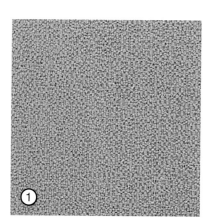

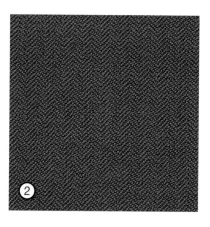

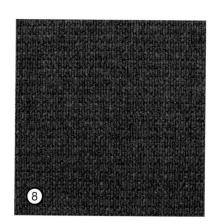

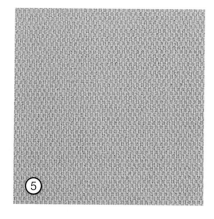

① Fawn
② Chocolate
③ Blue-grey
④ Pale sage
⑤ Slate
⑥ Sea grey
⑦ Pale blue
⑧ Red with blue stripes
⑨ Brown and gold
⑩ Brown
⑪ Wheat
⑫ Charcoal
⑬ Auburn
⑭ Gold
⑮ Blue
⑯ Mauve-brown
⑰ Straw
⑱ Cream
⑲ Ivory
⑳ Tan

Synthetics come in four basic types: acrylic, nylon, polyester, and poly-propylene (olefin), though these are often masked by proprietary trade names. Synthetic carpet may include some or all of these, and many carpets are made up of both wool and a synthetic. Cheaper, lightweight synthetic carpet can seem flat and tired after a few years, so it's worth spending more if you plan to keep the same floor for a long time.

① Vanilla
② Pepper
③ Dusky mauve
④ Dusky pink
⑤ Blue-grey
⑥ Sage
⑦ Blue slate
⑧ Beige
⑨ Espresso
⑩ Cappuccino
⑪ Indigo
⑫ Emerald
⑬ Straw
⑭ Earth
⑮ Barley
⑯ Seagrass
⑰ Deep rose
⑱ Lavender
⑲ Oat
⑳ Bronze

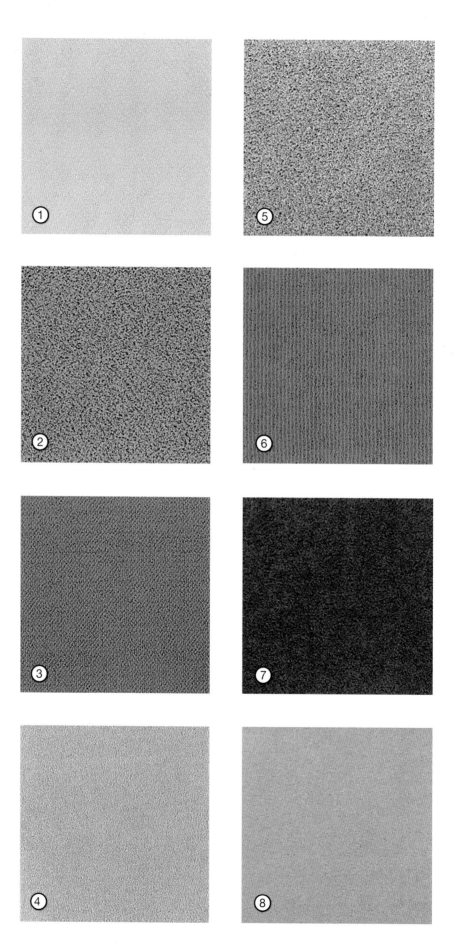

Tufted carpet makes up roughly nine tenths of all carpet sold. These are divided between cut-pile and loop-pile carpets, depending on how the yarn has been processed. Cut-pile carpet (as shown on these pages) is smooth and has a luxurious look and feel; common types are 'Saxony,' where the pile is cut to a level surface, and 'velvet,' which leaves a shorter pile that seems more formal and, possibly, less welcoming. A key point with cut-pile carpets is how much the yarn is twisted, as this protects it from damage. (See pages 208–209 for loop-pile examples.)

① Flowers
② Laurel diamonds
③ Lattice
④ Curved lattice
⑤ Traditional
⑥ Plants
⑦ Medieval
⑧ Renaissance
⑨ Fauna panels
⑩ Crocodile print
⑪ Zebra print
⑫ Autumn
⑬ Petal lattice
⑭ Python print
⑮ Antelope print
⑯ Gold-and-green grid
⑰ Interwoven garden
⑱ Cheetah print
⑲ Leopard print
⑳ Mosaic

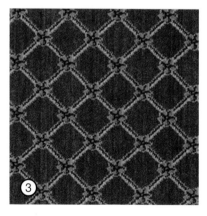

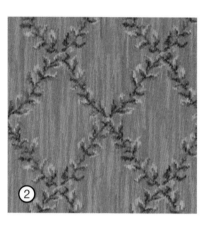

Loop-pile carpet is characterised
by the yarn being looped back
into the backing and then being
left uncut, which means that
there are no exposed yarn tips.
Consequently, this carpet tends
to last longer. It has more texture
than cut-pile versions, shows
footmarks and vacuum cleaner
lines less and is easier to clean.
Most loop-pile carpet is known
as 'berber' (after rugs made by
the African tribes of that name),
though there is a smoother
variety known as low-level loop.

① Contrasting lines
② Red check
③ Grey plaid
④ Jewels
⑤ Gold weave
⑥ Burgundy grid
⑦ Diamonds
⑧ Checked trellis
⑨ Stitched dash
⑩ Neutral stripes
⑪ Tufted pastel stripes
⑫ Tufted soft floral
⑬ Small check
⑭ Contrasting stripes
⑮ Tufted textured weave
⑯ Tufted two-tone vine
⑰ Diagonal check
⑱ Harmonising stripes
⑲ Tufted large check
⑳ Tufted blue stained-glass pattern

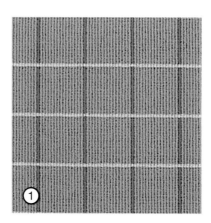
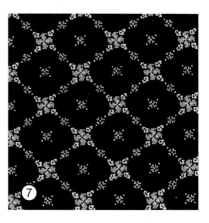
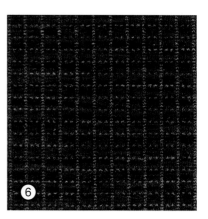
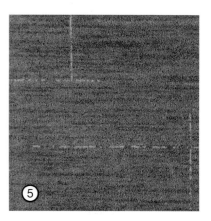
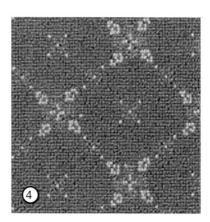
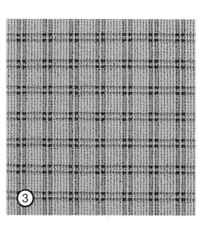

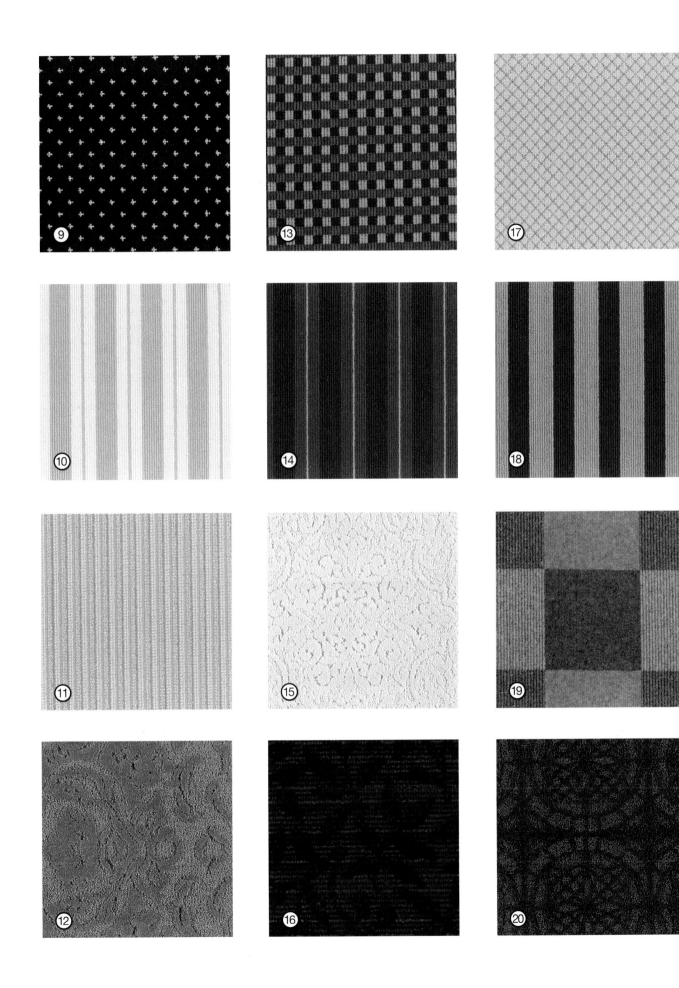

Natural Flooring

Natural flooring is increasingly popular for two reasons: its pared-down simplicity suits a lot of modern styling and it is ecologically sound because it is sustainably harvested. Natural fibres such as sisal, coir, jute, seagrass and rush offer mostly, but not exclusively, neutral tones and a rich texture. Interestingly, natural fibres shake off dirt with ease. The drawbacks are: some feel prickly, those such as seagrass can be slippery, and natural flooring is less durable than carpet, so it won't suit every application.

① Sisal is the most popular of the natural fibres because it is durable but fairly soft underfoot. Unlike many of its cousins it can be dyed, as this deep-brown example demonstrates.

② Jute is the softest of the natural fibres, making it suitable for low-use living rooms and bedrooms (but not hallways or family rooms).

③ Its 'dyability' means that sisal stains relatively easily. Protective sprays are available for neutral tones, like this off-white carpet.

④ Softer than seagrass and considered more elegant than sisal, abaca fibre has excellent water-resistance, making it a rising star in the world of natural flooring.

Sisal is the most popular natural fibre because it is so versatile in terms of function and appearance. It is harvested from a subtropical spiky bush and has been used for a long time to make rope and twine. Sisal yarn has an appealing, bold texture that is naturally sound-absorbing, antistatic and very durable. It is easy to dye, so it can take almost any colour, but is worth treating with an inhibitor afterwards to prevent staining.

① Stony grey
② Vanilla
③ Ochre
④ Fine straw
⑤ Cappuccino
⑥ Bamboo
⑦ Peanut
⑧ Coarse straw
⑨ Amber
⑩ Ivory
⑪ Medium hay
⑫ Charcoal
⑬ Slate
⑭ Coarse hay
⑮ Serge
⑯ Shaded cream
⑰ Copper
⑱ Fine hay
⑲ Diamond weave
⑳ Mocha

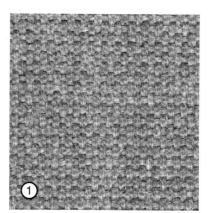
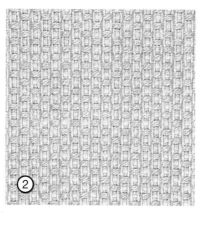
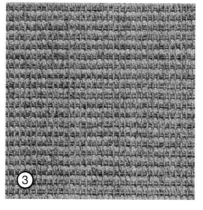
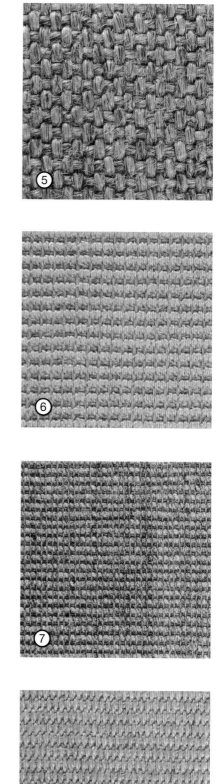

Continued from page 213.

1. Woven ammonite
2. Woven pyrite
3. Pearled brown steel
4. Pearled beech
5. Woven copper
6. Woven mahogany
7. Pearled African teak
8. Pearled straw
9. Pearled honeycomb
10. Braided wheat
11. Woven chamomile
12. Woven chestnut
13. Woven sand
14. Braided chocolate
15. Woven buff
16. Woven cinnamon
17. Woven biscuit
18. Braided pearl
19. Woven pebble
20. Woven pecan

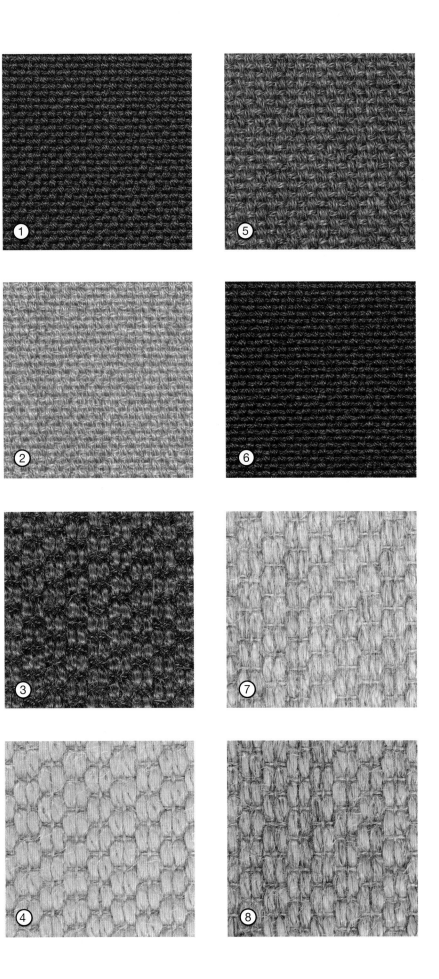

Abaca, or Manila hemp, is the strongest natural fibre so it is good for high-traffic areas.

Hemp tends to rot when it is exposed to water, so it must be used in well-ventilated rooms and, of course, cannot be dyed.

Jute is a soft fibre that is suitable for bedroom use but not for high-traffic areas.

Paper is an unusual carpeting material, made by twisting the pulp that is taken from softwoods.

Seagrass is the smoothest, most comfortable natural fibre.

① Wild rice abaca
② Buttermilk abaca
③ Chestnut abaca
④ Pearl abaca
⑤ Oatmeal abaca
⑥ Coffee abaca
⑦ Cream abaca
⑧ Bronze abaca
⑨ Coarse hemp
⑩ Braided jute
⑪ Palm paper
⑫ Interwoven seagrass
⑬ Interwoven hemp
⑭ Roped jute
⑮ Natural paper
⑯ Fawn seagrass
⑰ Natural jute
⑱ Braided paper
⑲ Grey seagrass
⑳ Natural seagrass

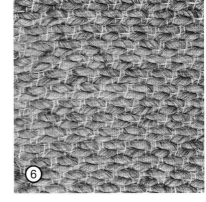

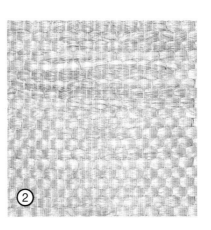

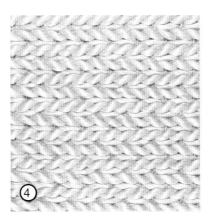

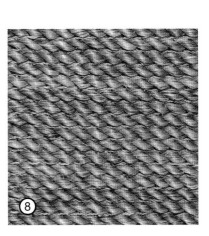

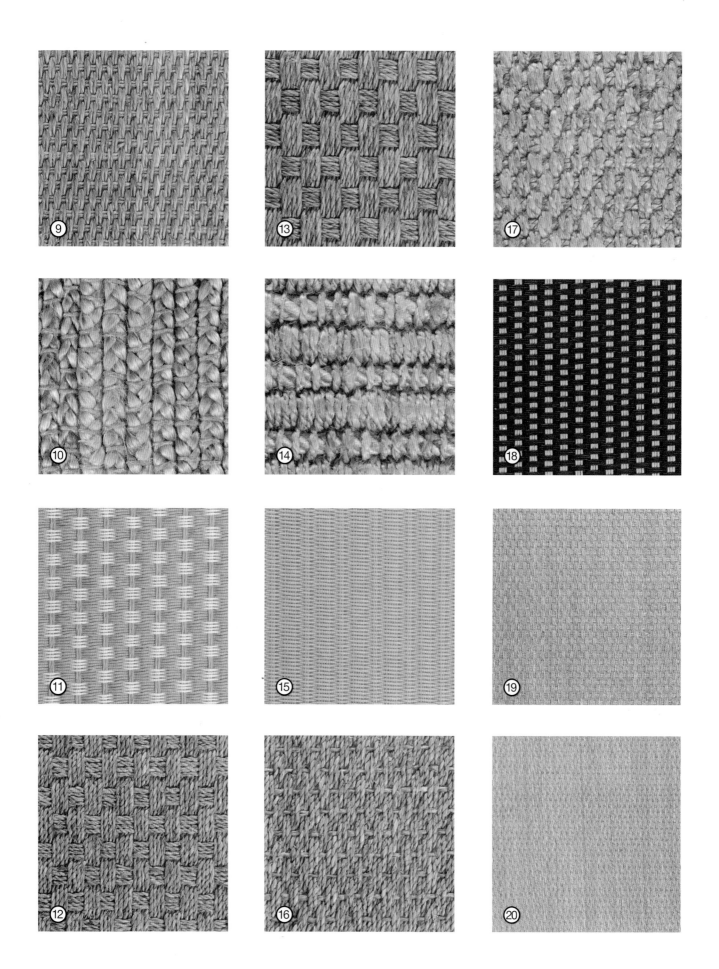

Rugs

You can transform a room just by laying a rug on the floor. These floor coverings add colour, texture, comfort, warmth and sound protection. This means they can complement any other type of flooring by countering its disadvantages. For example, rugs bring warmth to stone, protect a wooden floor from dirt and scratches, lessen noise and bring a touch of domesticity to an austere tile surface. They'll even hide a stain until you can remove it! Their only drawback is that they can slip if not properly adhered to the floor. Lay rugs over a slip-resistant mat and secure the edges.

① This Oriental-style rug reflects the classical symmetry of its setting.

② A large rug is as good as carpet for noise insulation, comfort and protecting a wooden floor.

③ This foliage design brings the room to life.

④ Abstract designs are ideal for contemporary décor.

⑤ This rug's curved shape counterbalances the severe lines of the window frame and furniture.

⑥ Rugs can add texture, bringing a sense of warmth to what might otherwise be a cool, unwelcoming entrance hall.

Rugs have been woven for centuries, with different regions often using the same styles and patterns for generations. Oriental-style rugs mimic designs that originated in western and Central Asia, North Africa and southeastern Europe. In these areas rug-making is a traditional art form. Many traditional weaves and styles have been given modern interpretations, with new colours or modern motifs. Consider texture as well as colour and pattern. Does the room call for a simple flat weave or a sumptuous thick pile?

① Hexagons
② Circles
③ Sage fans
④ Gold frame
⑤ Swirling leaves
⑥ Autumn leaves
⑦ Formal garden
⑧ Art deco
⑨ Diamonds
⑩ Floral tapestry

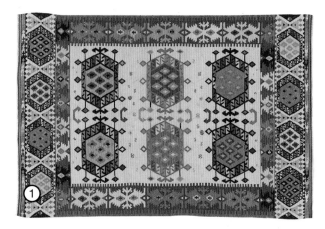

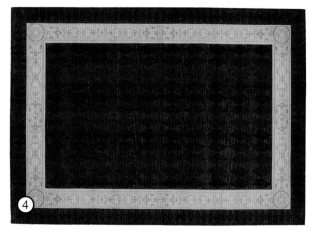

Continued from page 221.

① Oval
② Summer garden
③ Plaster relief
④ Abstract
⑤ Celtic knots
⑥ Chiesa Yellow by Suzanne Sharp
⑦ Folia by Emily Todhunter
⑧ Star-shaped leaves
⑨ Faded Glory Red by Paul Smith
⑩ Door by Committee

Contemporary rugs are available in a huge range of subtle colours, delicate patterns and interesting textures, as these close-ups demonstrate. Traditional Oriental-style rugs tend to dominate the market, but in the wrong surroundings they can seem merely functional. Seek out rugs that match the style of your décor. There are rugs for all settings!

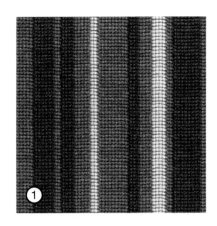
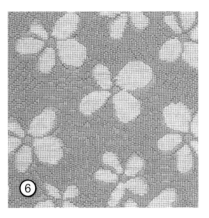
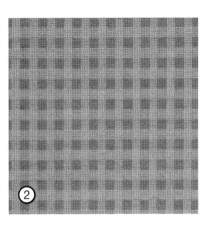
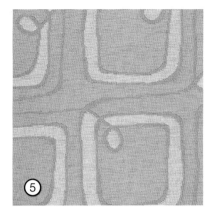
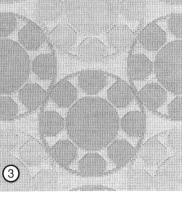

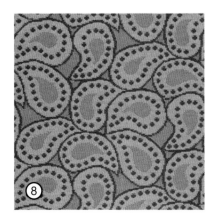
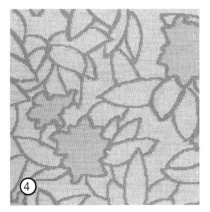

① Stripes
② Plaid
③ Wheels
④ Leaf-and-flower design
⑤ Abstract
⑥ Daisies
⑦ Roses
⑧ Contemporary paisley
⑨ Textured grey
⑩ Red-and-brown stripe
⑪ Fine chalk
⑫ Textured mushroom
⑬ Textured green
⑭ Lime-and-grey stripe
⑮ Deep pile seaweed
⑯ Textured white
⑰ Textured cream
⑱ Royal red knit
⑲ Black-and-grey baubles
⑳ Brown knit

Using Colour

The colour of the floor often sets the tone for the whole room, and soft flooring materials offer innumerable choices. One thing to consider is the period of your home. Will the palette associated with that era enhance the décor today? For example, Shaker décor combines blue or red with cream and biscuit shades. Victorian colour schemes tend to be darker and stronger, such as olive green and walnut. This approach takes you away from the contemporary beige and cream palette (which essentially creates a blank canvas) and prompts fresh inspiration. Of course, you can add period colours to a modern house if you want to.

① Pastels often complement each other, as with this pale blue carpet and the soft yellows of the curtains.

② Deep blue and cream is a classic combination from the Victorian period, though it looks thoroughly contemporary here.

③ An aqua carpet brings out the green in the motifs on the curtains, creating a fresh and natural look.

④ A red carpet brings out the gold hues in the curtains, making the room seem more formal and grand.

⑤ Create a Mediterranean feel by combining orange with fresh turquoise.

⑥ The colours in this rug bring together the orange-brown carpet and the beige sofa. Notice how the neutral walls virtually 'disappear'.

⑦ Orange, yellow and brown create a classic 1970s look that is now fashionably retro.

⑧ The warmth of the brown in this rug has been brought out with bright red and purple accents.

Practicalities

Floors do a tough job and they need to be well maintained in order to help them stay beautiful and effective for as long as possible. Consider how much time, effort and money you are prepared to invest in this before you finalise your choice of material: some fantastic floors are very high maintenance (especially leather!). Hard floors will need to be swept daily and washed weekly. Soft floors need regular vacuuming and occasional professional cleaning.

Identify the high-traffic areas of your home and plan the flooring accordingly. Entrances, staircases, landings and kitchens take a lot of punishment from busy feet, so go for the most durable material you can afford. In living rooms and bedrooms comfort is the priority, so plush carpet or thick-pile rugs are excellent choices.

FLOOR MAINTENANCE

Regular care will prolong the life of the floor and help ensure that warranties are honoured. Dirt and stains are inevitable, and the longer they are left, the more damage they will do. The first step in keeping all of your floors in good condition is to keep the dirt out. Don't have sand or small rock particles near the door entrance because they will find their way in. Put dirt-trapping mats at external doors – that means coconut matting, not carpet. Lay a good hallway carpet which,

SOUND FLOORING ADVICE

Noise is a factor that is often overlooked in the hunt for flooring, but once you are living with too much, it can have a major impact on your quality of life. Hard floors are inevitably noisier than soft surfaces, and in a room that echoes (such as a kitchen with a high ceiling), it is worth considering whether stone or wood will be too hard on your ears. Similarly, the sound of pacing along an upstairs corridor, just above your calm oasis of a living room, can be very irritating. Carefully consider whether you want wood in upper stories of the house, and if so what noise insulation could be laid underneath it.

according to an old saying, should be 'four lazy steps long' (about 15 feet or 4.5 metres). This allows time for grit and dirt to drop off shoes before it can get into your home. If you only have space for a doormat, fold a few sheets of newspaper

BROOM AT THE TOP

Clean brooms work better, so keep your broom clean using a mild detergent and warm water. If the broom has real bristles, they will work better if you dip the head in salted water before use. Store brooms head up, handle down to keep them in good condition.

under it to collect the dirt – this is much easier to remove than dirt that is ground into your carpet.

STONE

Many people do not realise that most stone (especially limestone and sandstone) is porous and will absorb water and stains. This means that it must be sealed when used as a flooring material and will require further applications as the floor undergoes the wear and tear of everyday life. This sealant will alter the appearance of the floor, so always seek advice on how to apply it. Do not polish smooth-textured stone, such as marble, because it will make your floor as slippery as an ice rink. Brick can be polished, sealed (except for absorbent types), or covered with linseed oil.

MAINTAIN GROUT APPEARANCE

In any stone or tile application where the grout will be visible, it is the weakest link in terms of maintenance and appearance. Dirt gathers in the depression, and grout itself can wear and discolour. Avoid wide grout joints, as they increase the problem.

Sweep or vacuum stone floors daily and wash them weekly in mildly soapy water. Make sure you dry them with a floor cloth to get rid of excess water and to stop salts from seeping out from the stone, which could create marks. Wash terrazzo floors with a little scouring powder, then rinse it with clean water. Avoid using wax polishes because they will make the floor slippery.

HARD FLOORS

Porous tiles, such as the terracotta, quarry and encaustic (inlaid) varieties, must be sealed (and resealed). Use a commercial sealant or the long-used recipe of equal parts beeswax and boiled linseed

CLEVER DUSTING

Impregnated floor-dusting cloths attract dust and give the floor a light shine. You can make your own using squares of an old woollen blanket soaked in equal quantities of paraffin (or kerosene) and malt vinegar. Hang it up to dry and store in an airtight bag when not in use.

oil (some people prefer to increase the proportion of linseed oil). Mix, cool, apply and then buff. Sometimes these floors release salt deposits, leaving a white patch. This can simply be washed away with clean water.

Other hard-tile floors may need occasional resealing, but they can be cleaned with mild household cleaning products and don't need specialist products. Mop up spills quickly though, because acidic liquids like alcohol and citrus juices can stain, especially on unsealed tiles. If this happens, try removing the mark with mineral spirits.

Concrete floors have to be sealed to stop the surface from dusting. After that, as with metal and glass, a gentle detergent and hot water is all you need.

Floor movement can break down the adhesive layer beneath ceramic tiles, causing major cracks in the grout. The tiles themselves are also vulnerable to damage. It makes sense to keep back a few spare tiles that can be used as replacements. Always check the underlayer to make sure it is smooth, solid and level before laying the substitute tile.

WOOD

Wood is a fairly soft material compared with stone, so it can be dented by heavy objects and stiletto heels. Heels can cause enormous damage – permanent damage in the case of wood-veneer flooring. If subjected to too much water, wood will swell or even warp, so you have to be very careful with your cleaning regime. Prefinished wood floors are easy to maintain because the sealer is baked into the wood. Most urethane and Swedish finishes can be damp-mopped with white vinegar and water. Mix 250 mls vinegar for every 4 litres of water (one cup of vinegar per gallon of water), or use a mild detergent.

Untreated boards must be varnished, oiled or waxed – a process that will need to be repeated regularly (as often as every six months in some environments). When cleaning, beware of using oil soaps, lemon oil, sprays, liquid waxes or other products, and always check with the manufacturer or supplier before applying them. You can judge if it's time to reseal wood floors by checking the finish. If the finish has become dull or if water soaks in, rather than resting on the surface, it is time to seal again.

The cleaning routine is simple: sweep the floor frequently with a soft broom to remove abrasive particles (unswept grit will get trodden in and gouge the top layer) and mop up spills quickly so that water doesn't soak into joints or the wood itself. Wash with mild, diluted detergent once a week, being careful not to get the floor overly wet, and dry it off with a dry mop straight afterwards.

Protect the surface of a wooden floor by attaching felt pads to the bottom of the furniture. Cover chair legs with protective rubber caps to

SILENCE THE SQUEAKS AND CREAKS

Wooden floors move over time, and older wooden floors are liable to develop creaks and squeaks. Here are a few tips to deal with them:

- Tap slivers of wood into gaps between floorboards or fill wider gaps with papier-mâché.
- Add wood dye to a filler to camouflage your work.
- Minor blemishes can be treated with coloured putty. If the floor has a variety of tones, put a different colour on each finger so that you have one to match.
- Silence creaks by sprinkling graphite powder or talcum powder into the joints.
- If floorboards are creaking, nail them – at an angle, if necessary – into the joist below.
- Warped boards should be screwed in rather than nailed down because the screw holds them more firmly.

prevent scraping and scratching. If you have chairs with plastic castors (like office chairs), you can buy clear plastic mats to protect the wooden floor. Rubber wheels do not need further protection.

A potted plant can be a swift route to a damaged floor. Place it on half an inch (1 cm) of dense cork so that no water seeps through and check the overflow bowls regularly.

RESILIENT FLOORS

This type of flooring is most susceptible to damage from the wrong kind of treatment. Vinyl has a reputation as an easy-maintenance material, but while it is resistant to water, oils, fats and many chemicals, it will be damaged by sharp or abrasive objects such as grit and stiletto heels. In addition, vinyl can be damaged by cleaning products such as bleach, scouring powder and strong alkaline detergents.

Treat as recommended for wood, sweeping it regularly, wiping away spills promptly, and cleaning with a mild, diluted detergent.

Linoleum should be treated similarly and will resist most stains, apart from those with high solvent content such as dry-cleaning fluid, nail polish and oven cleaner. Linoleum has been used for a long time. One time-honored recipe for refreshing it is to wipe it down with one part fresh milk mixed with one part turpentine: rub the mixture in and polish with a warm, soft cloth. Almost as old is the problem of children's wax crayons marking places they shouldn't. If they mark vinyl or linoleum, a little silver polish will remove it.

Three natural materials in this category of flooring are rubber, leather and cork. Fats and solvents will mark rubber, so wipe away spills quickly. When you've mopped a rubber floor, go over it again with a clean, damp mop to get rid of any detergent residue.

Leather is more resistant, as long as it is waxed regularly (three times before use and then about twice a year), but scratches are inevitable. Indeed, they will become part of the character of the floor, a sort of patina of life.

Cork needs a lot of care. It must be sealed to make it resistant to water and stains. Even presealed cork should be given an extra coat after installation so that the joints are protected. If the surface becomes marked or soiled, light sanding will remove the problem, but you may need to renew the seal or polish. This must always be done, in any case, as soon as there are signs of wear and tear.

MOVING HEAVY OBJECTS

One definite way to damage a floor is to drag a heavy object across it. That is exactly what is required sometimes, of course, when moving a stove or sofa. Here are a handful of suggestions:

- Place a piece of carpet under the object so that you can simply slide it along.
- Protect the floor along the route with a piece of plywood.
- When moving furniture across a carpet, put a foil pie dish under each leg and slide it to where you want it.

SOFT FLOORS

When a carpet is first laid, it will shed pieces of fluff for a few weeks, and it is advisable to avoid vacuuming during this period. After that, vacuuming weekly is essential. This is because dirt that is left on the floor will collect and get crushed together, matting the fibres and making the carpet look tired and old, even if the pile is not very worn. It is also worth having carpets and rugs thoroughly cleaned every so often, because they trap and filter contaminants. A thorough cleaning will remove these unwelcome visitors from your house, making it a healthier environment. When shampooing carpet, remove all furniture from the room to ensure an even treatment on all of the flooring.

Carpets and rugs need careful handling because, like all textiles, they stain easily. You can have your carpet sprayed with stain repellent, but this won't solve the problem completely. If or when the worst happens and there is a bad spill, act quickly. Over time, stains chemically react with the carpet and this makes them harder to remove. Use a blunt knife to scrape away any solid mess immediately. Do not add any liquid because this will only spread the mark. For similar reasons, blot liquid spills with a damp cloth or a paper towel so that as little as possible is absorbed into the pile. When mopping, work from the outside in to avoid spreading the problem. Also avoid wetting the carpet too much, as this can cause further damage.

To effectively tackle a stain it is important to know what caused it:

water-based stains can be removed using warm water and a little mild detergent (such as carpet shampoo). Oil-based stains require dry-cleaning fluid or solvent applied with a damp cloth. Some stains, such as mud, can be rubbed or vacuumed away once they have dried.

FURNITURE DENTS

Heavy furniture will dent a carpet – not a problem until you rearrange the room! The simplest answer is to use a coin to rub the dent so that the pile stands up. If this fails, try an ice cube: as it melts, the water swells the fibres. Running over the area with a vacuum cleaner afterwards lifts the pile. To avoid another dent, buy carpet-protector dishes. Alternatively, you may find that jam or coffee-jar lids do the job.

TRADITIONAL CARPET-CLEANING TECHNIQUES

Snow and tea are the carpet cleaner's friends! Persian carpets used to be beaten to get rid of dust and then laid facedown in fresh snow. The snow would draw out any leftover dirt and freshen the colours. In Georgian England, damp tea leaves were mixed with lavender heads and sprinkled over rugs. After it had absorbed dirt and odors, the mixture was dried off. A time-tested method that can still be used today is to freshen carpets by sprinkling them with salt, oatmeal, or corn flour. Leave for a couple of hours and then vacuum. Try reviving faded carpet with a mix of one part vinegar to two parts boiling water. Soak a cloth in the solution and then rub it into the carpet.

With soft floors, you must consider wear. Wear patterns vary depending on which areas get the heaviest traffic. You can protect and conceal a worn area with a rug, making sure that it is secured to the floor with double-sided tape for safety. When the carpeting on stairs starts to wear, take it up and lay it again in reverse, top to bottom, so that the treads become risers. This will double the life of the carpet.

NATURAL FIBRES

Natural-fibre flooring materials (sisal, jute, coir and seagrass, for example) are very sensitive to water and should never be washed or shampooed. If natural flooring does get wet it will either swell or shrink and could start to rot. The exception to this rule is rush matting, which rots if it dries. To avoid deterioration dampen it weekly with a sprayer. Vacuum natural fibres regularly and try to brush out any marks as soon as you notice them. As natural fibres are harder and static-free, they will attract less bacteria and less dust than synthetic fibres.

Glossary

Adhesive: The bonding agent joining the floor covering to the underlayer.

Backing: The underside of the carpet that touches the subfloor.

Baseboard: Moulding at the bottom of a wall that covers the joint with the floor.

Basketweave: A checked, woven pattern created by laying tiles or bricks in a pattern.

Beam: A horizontal framing object, such as a joist or header.

Berber: A type of loop-pile carpet.

Binding: Fabric at the edge of carpets or rugs to prevent unravelling.

Border: A decorative strip in a contrasting colour often laid around the edge of a room.

Carpet bar: see Threshold.

Ceramic: The stonelike material made by firing clay at high temperatures.

Clear finish: A wood finish that does not conceal the grain or discolour the wood.

Click system: An installation method for wooden floors in which the tongues and grooves are slotted together without the need for glue.

Combination floor: A floor that consists of two or more materials. Also known as mixed-media flooring.

Cushion: See Padding.

Cut and loop: A type of carpet pile with both intact and cut yarn loops. This combination produces a more textured finish.

Cut pile: The carpet-making method where the loops are all cut, also known as 'shag' or 'shag pile'.

Damp-proofing: A sheet or membrane laid to prevent moisture from rising up into the floor.

Density: Method of grading carpet according to the amount of pile yarn or fibre used and how close together the fibres are.

Draughtboard: A pattern of alternating, contrasting colours.

Embossing: A process to add texture to stone or carpet.

Encaustic: A type of tile inlaid with another colour during the firing process.

Engineered flooring: Also known as structured flooring, this comprises several layers of wood glued together into strips or planks to look like solid wood, offering a cheaper alternative to solid wood.

Faux finish: A decorative effect to change the appearance of one material into that of another (e.g., wood to look like stone).

Flagstone: Quarried stone slabs that are up to 8 cm thick, mostly used outdoors.

Flat-weave: A woven carpet without a pile.

Floating floor: Wood or laminate floor that isn't attached to the subfloor, but rests instead on a thin foam padding.

Glaze: A decorative and water-proof coating used to cover tiles.

Grout: The material inserted between tiles and at the edges of rooms to seal gaps.

Hard twist: A type of cut pile where the fibres are twisted to give a tight, durable texture.

Hardwood: The wood from broad-leafed or deciduous trees.

Herringbone: A fishbone pattern of rows that slope in opposite directions.

Inlay: A contrasting material set in pieces into a surface to form a design.

Insert: A small, decorative tile or baton used to create an accent or pattern.

Joist: The framing objects that support the floor.

Laminate: A simulated floor effect made by attaching a plastic layer to high-density fibreboard.

Levelling compound: A blend of liquid latex and powder, used to level a concrete subfloor.

Linoleum: Flooring material made from linseed oil, wood and cork powder, pine resin, ground limestone and jute, forming a biodegradable material with antistatic and hypoallergenic qualities.

Longstrip flooring: A material that is made by fastening up to three strips of wood together to form each plank.

Matte: A dull finish.

Medallion: A large, decorative wood or tile design.

Moulding: The decorative wooden strip along walls and floors.

Nonwoven carpet: The cheapest kind of carpet, made by bonding synthetic fibres to a backing.

Oriental rug: A handwoven or handknotted rug from the Middle East or Far East.

Padding: The material laid under a carpet to improve its performance. Also known as a carpet cushion.

Paisley: The comma-shaped motif that is occasionally used in traditional carpets.

Parquet: Flooring comprised of small blocks of wood laid to create patterns on the floor.

Paver: A brick produced for use in flooring.

Pile: Visible carpet fibre, also known as 'face' or 'nap'.

Plaid: A pattern made up of crossed stripes.

Plush: A velvet-smooth carpet finish.

Ply: Single fibre yarn.

Plywood: A basic material commonly used as an underlayer for resilient and ceramic tile floors.

Prefinished: Wood or laminate board supplied with a protective coating.

Radiant heating: Underfloor heating using pipes or mesh netting.

Raised floor: A type of floor used in offices in which wiring, cables and pipes can be easily accessed beneath. Also known as a suspended floor.

Reclaimed wood: Wood that has been used for another application; e.g., railway sleepers.

Reproduction stone: Manufactured concrete stone that imitates natural materials.

Resin: A treatment poured over concrete to form a hard floor. The end result is similar to polished concrete or poured rubber.

Rise: The height of one step in a flights of stairs.

Satin: A soft sheen finish.

Saxony: Cut-pile carpet with a very smooth surface.

Sealant: A product used to protect flooring surfaces from stains and water damage.

Serape: A brightly coloured, woven Mexican rug.

Shag pile: A long carpet pile of up to 5 cm (2 inches).

Sheet vinyl: Plastic flooring in 2, 3 or 4 metre-wide sheets.

Slab: A lengthwise cut of a large quarry block of stone.

Solid wood floor: Wooden floor that is made of wood all the way through (i.e., not in layers).

Spacer: Installation accessory that is placed between tiles to regulate gaps.

Stencilling: A decorative paint technique in which a motif is repeated with the use of a template.

Strip floor: Wood or laminate flooring less than 10 cm (4 inches) wide.

Subfloor: The layer, often made of plywood boards or concrete, that lies on top of the joists and underneath the flooring surface.

Surface layer: See Wear layer.

Tackless strips: Wood or metal strips with tacks onto which carpet is stretched and secured. They are called 'tackless' because they eliminate the need for tacking carpet from above.

Tesserae: The individual pieces of a mosaic, often attached to a backing to form a tile.

Threshold: The transition strip between two rooms or two flooring types.

Tongue and groove: A system of protuberances and slits on strips of wood that can be connected to attach each side to the other. Used in the 'click' system.

Traffic: The amount of footfalls along the same path.

Trim moulding: A decorative strip of wood attached to a baseboard.

Tufted carpet: A carpet made by feeding yarn through a plastic backing sheet.

Tumbling: A literal tumbling process for ageing the look of new stone.

Twist: The process of winding carpet fibre around itself to increase its strength.

Underlayer: Any layer positioned between the subfloor and the floor covering.

Urethane: An artificial form of wax used for surface finishing.

Veining: The pattern created by the different natural colours found in stone.

Waterproofing membrane: See Damp-proofing.

Wear layer: The top layer of a multilayered flooring material, often designed to protect the decorative layer below.

Wool: A natural animal fibre used for carpets.

Woven carpet: Carpet made by weaving yarn on a loom to form a fabric with interlocking warp (width) and weft (length) yarns.

Yarn: A continuous strand of fibres.

Resource Guide

The following list of manufacturers, associations, and outlets is meant to be a general guide to additional industry and product-related sources. It is not intended as a complete listing of products and manufacturers represented in this book.

ASSOCIATIONS

Concrete Network
www.concretenetwork.com

National Institute of Carpet and Floor Layers
Promotes excellence in floor-laying, provides list of members
4d St Mary's Place
The Lace Market
Nottingham NG1 1PH
Tel: 0115 958 3077
www.nicfltd.org.uk

The Tile Association
Promotes high standards in floor and wall tiling, provides list of members
Forum Court
83 Copers Cope Road
Beckenham
Kent BR3 1NR
Tel: 020 8663 0946
www.tiles.org.uk

STONE FLOORS

Bettini Tile Service
Tel: +39 516242166
www.bettini.com

Indigenous Ltd
Cheltenham Road, Burford
Oxfordshire OX18 4JA
Tel: 01993 824200
www.indigenoustiles.com

Mandarin Stone
Unit 1, Wonastow Road Ind Estate
Monmouth
Monmouthshire NP25 5JB
Tel: 01600 715444
www.mandarinstone.com

HARD FLOORS

American Olean/Daltile
tiles
7834 Hawn Fwy
Dallas, TX 75217
USA
Tel: +1 214 398 1411
www.americanolean.com

Artwork in Architectural Glass Studios
glass floors
20101 SW Birch
Suite 276
Newport Beach
CA 92660
USA
Tel: +1 949 251 0075
www.aag-glass.com

Azurra Mosaics
mosaic tiles
P.O. Box 2801
Purley CR8 1WX
Tel: 0845 090 8110
www.mosaics.co.uk

Carina Works
metal tiles & planks
12934 Nutty Brown Rd
Austin, TX 78737
USA
Tel: +1 800 504 5095
www.carinaworks.com

Dynamic Stone
terrazzo
6039 Ontario Street
Vancouver
British Columbia V5W 2M3
Canada
Tel: +1 604 328 4777
www.dynamicstoneinc.com

Fired Earth
Tiles
3 Twyford Mill, Oxford Road
Adderbury, Nr Banbury
Oxfordshire OX17 3SX
Tel: 0845 366 0400
www.firedearth.com

Florida Tile
tiles
P.O. Box 447, Lakeland
FL 33802 USA
Tel: +1 800 352 8453
www.floridatile.com

Seattle Glass Block
glass floors
6029 238th Street SE
Woodinville, WA 98072 USA
Tel: +1 800 829 9419
www.seattleglassblock.com

Topps tiles
Tiles and wood flooring
Oak Green Business Park
Earl Road, Cheadle Hulme
Cheshire SK8 6QL
Tel: 0800 783 6262
www.toppstiles.co.uk

WOOD FLOORS

Anderson Floors
P.O. Box 1155
Clinton, SC 29325 USA
Tel: +1 864 833 6250
www.andersonfloors.com

Award Hardwood Floors
401 N. 72nd Avenue
Wausau, WI 54401 USA
Tel: +1 888 862 9273
www.awardfloors.com

The Hardwood Flooring Company Ltd
31–35 Fortune Green Road
London NW6 1DU
Tel: 020 7431 7000
www.hardwoodflooringcompany.com

Kährs UK Ltd
Unit 2 West, 68 Bognor Road
Chichester, West Sussex PO19 8NS
Tel: 01243 778747
www.kahrs.com/uk

Lauzon
2101 Côte des Cascades
Papineauville, Quebec J0V 1R0
Canada
Tel: +1 877 427 5144
www.lauzonltd.com

Mullican Flooring
815 Love Street, P.O. Box 3549
Johnson City, TN 37604
USA
Tel: +1 800 844 6356
www.mullicanflooring.com

Pergo
PO Box 13113, Kingsbury Link
Piccadilly, Tamworth B77 9DJ
Tel: 01827 871841
www.pergo.com

Quick-Step
Ooigemstraat 3
8710 Wielsbeke
Belgium
Tel: +32 56 675211
www.quick-step.com/europe

SW Advanced Flooring
Sells Lauzon and Kahrs flooring
Unit 6B1 Dalhousie Business Park
Bonnyrigg EH19 3HY
Tel: 0131 663 5802
www.swadvancedflooring.co.uk

Teragren
12715 Miller Road NE, Suite 301
Bainbridge Island, WA 98110
USA
Tel: +1 800 929 6333
www.teragren.com

RESILIENT FLOORS

Amorim
cork
Rua de Meladas, N° 380
Apartado 20, 4536-902 Mozelos
Portugal
Tel: +351 227 475 400
www.amorim.com

Amtico
vinyl
Solar Park, Southside, Solihull
West Midlands B90 4SH
Tel: 0800 667766
www.amtico.com

Congoleum
vinyl and tiles
P.O. Box 3127, Mercerville
NJ 08619-0127
USA
Tel: +1 800 274 3266
www.congoleum.com

Duro Design
cork and wood
2866 Daniel-Johnson Boulevard
Laval, QC H7P 5Z7
Canada
Tel: +1 888 528 8518
www.duro-design.com

Forbo Nairn Ltd
linoleum
P.O. Box 1, Kirkcaldy, Fife KY1 2SB
Tel: 0800 731 2369
www.forbo-flooring.co.uk

Interior Leather Surfaces
leather
5 Lakeshore Close, Sleepy Hollow
NY 10591 USA
Tel: +1 877 231 2100
www.leathertile.com
www.interiorsurfaces.com

SOFT FLOORS

Axminster Carpets Ltd
Axminster, Devon EX13 5PQ
Tel: 01297 630650
www.axminster-carpets.co.uk

Brintons Limited
PO Box 16, Exchange Street
Kidderminster
Worcestershire DY10 1AG
Tel: 0800 505055
www.brintons.net/

GuildCraft Carpets
105 East 5th Street, Suite 200
Northfield, MN 55057
USA
Tel: +1 507 664 9500
www.guildcraftcarpets.com

Karastan
508 East Morris Street
Dalton, GA 30721
USA
Tel: +1 800 234 1120
www.karastan.com

Merida Meridian
643 Summer Street
Boston, MA 02210
USA
Tel: +1 800 345 2200
www.meridameridian.com

Nourison Europe
Freilagerstr. 47
Postfach 135
8043 Zurich
Switzerland
Tel: +41 1 401 4546
www.nourison.com

The Alternative Flooring Company
3b Stephenson Close, East Portway
Andover, Hampshire SP10 3RU
Tel: 01264 335111
www.alternative-flooring.co.uk

The Rug Company
124 Holland Park Avenue
London W11 4UE
Tel: 020 7229 5148
www.therugcompany.org

Wilton Carpet Factory Shop
King Street, Wilton, Salisbury
Wiltshire SP2 0AY
Tel: 0845 296 8850
www.carpetweaver.co.uk

GENERAL

Armstrong
linoleum, vinyl, wood, laminate, tiles
Hitching Court
Abingdon Business Park
Abingdon, Oxfordshire OX14 1RB
Tel: 01235 554848
www.armstrong-europe.com

Mannington
vinyl, wood, laminate, tiles, carpet
75 Mannington Mills Road
Salem, NJ 08079
USA
Tel: +1 800 482 0466
www.mannington.com

Index

The publishers would like to thank the following companies for their invaluable assistance: Amorim, Anderson Floors, Armstrong, Artwork in Architectural Glass Studios, Award Hardwood Floors, Bettini Tile Service, Carina Works, Congoleum, Daltile, Duro Design, Dynamic Stone, Florida Tile, Forbo, GuildCraft Carpets, Indigenous, Interior Leather Surfaces, Kährs, Lauzon, Mandarin Stone, Mannington, Merida Meridian, Mullican Flooring, Nourison, Pergo, Quick-Step, The Rug Company, Seattle Glass Block and Teragren.

2 Kährs
4–5 Mandarin Stone
6 Kährs
7 Merida Meridian
8–9 1 Nourison; 2 Pergo; 3 Mullican Flooring; 4 Concrete Network
10–11 1 Congoleum; 2 Merida Meridian; 3 © Rob Marmion/Fotolia.com; 4 Kährs; 5 Forbo
12–13 1 Forbo; 2 Kährs; 3 Peter Woloszynski/ Redcover.com; 4 Ken Hayden/Redcover. com Architect: Michael Wolfson; 5 Keith Scott Morton/Redcover.com; 6 Andreas von Einsiedel/Legorreta & Legorreta Architects
14–15 1 Colin Sharp/Redcover.com; 2 Fabio Lombrici /Redcover.com; 3 Andreas von Einsiedel/Designer: Wessel van Loringhoven; 4 Mullican Flooring; 5 Guglielmo Galvin/ Redcover.com/Designer: Paul Warren; 6 Quick-Step
16–17 1 Mullican Flooring; 2 David Livingston/ Redcover.com; 3 Congoleum; 4 Alun Callender/ Redcover.com; 5 Forbo
18–19 1 Artwork in Architectural Glass Studios; 2 Nina Assam/Redcover.com; 3 Henry Wilson/ Redcover.com; 4 Andreas von Einsiedel/ Designer: Lester Bennett; 5 Florida Tile
20–21 1 Mandarin Stone; 2 Merida Meridian; 3 Artwork in Architectural Glass Studios; 4 Mullican Flooring; 5 Congoleum; 6 Forbo
22–23 1 Mullican Flooring; 2 Richard Powers/ Redcover.com; 3 Richard Holt/Redcover.com; 4 Niall McDiarmid/Redcover.com; 5 Kährs
24–25 1 Kim Sayer/Redcover.com; 2 Ken Hayden/ Redcover.com/Architect: Michael Wolfson; 3 Quick-Step; 4 Mullican Flooring; 5 Henry Wilson/Redcover.com
26–27 1 Anderson Floors; 2 Forbo; 3 Mannington; 4 Daltile; 5 Andreas von Einsiedel/Emily Todhunter
28 Andreas von Einsiedel/Designer: Natalie de Vilmorin
29 Daltile
30–31 1 Andreas von Einsiedel/Designer: Candy & Candy; 2 Andreas von Einsiedel/Designer: Michael Palyford; 3, 4 Indigenous; 5 Daltile
32 1 Mannington
34–35 1 Grey Crawford/Redcover.com; 2 Daltile; 3 Jake Fitzjones/Redcover.com; 4 Alex Smith/ Redcover.com
36–37 Daltile
38–39 1, 5 Mandarin Stone; 2–4, 6 Indigenous
40–41 Mandarin Stone
42–43 1 © Bruce Edward/Corbis; 2 Di Lewis/ Redcover.com; 3 © Scott Van Dyke/Beateworks/ Corbis; 4 Indigenous; 5 Simon McBride/ Redcover.com
44–45 Daltile
46–47 1, 5 Andreas von Einsiedel/Designer: Candy & Candy; 2–4, 6 Indigenous
48–49 Daltile
50–51 1, 4 Indigenous; 2 Johnny Bouchier/ Redcover.com; 3 Andreas von Einsiedel/ Designer: Smiros & Smiros Architects
52–53 Bettini Tile Service

54–55 1–3, 5 Indigenous; 4 Bieke Claessens/ Redcover.com
56–57 1–3, 5, 6, 9–11, 16, 18–20 Indigenous; 4, 7, 8, 13–15 Daltile; 12, 17 Mandarin Stone
58–59 1 Andreas von Einsiedel/Designer: Avril Giacobbi; 2 Daltile; 3 Wayne Vincent/Redcover. com; 4 Indigenous
60–67 Daltile
68–69 1, 3 Dynamic Stone; 2, 4 Beateworks Inc./ Alamy; 5 David Giles/EWA Stock
70–71 National Terrazzo & Mosaic Association
72–73 Indigenous
74 Daltile
75 Concrete Network
76–77 1 Indigenous; 2, 4 Daltile; 3 Ed Reeve/ Redcover.com; 5 Concrete Network
78–79 1 Andreas von Einsiedel/Designer: Ou Baholyodin; 2–6 Daltile
80–85 Daltile
86–89 Concrete Network
90–91 1 Andreas von Einsiedel/Designer: Isabelle de Borchgrave; 2 Richard Holt/Redcover.com; 3 Simon McBride/Redcover.com; 4 Winfried Heinze/Redcover.com
94–95 1 James Mitchell/Redcover.com; 2 James Mitchell/Redcover.com; 3–5 Seattle Glass Block; 6 Daltile
96–99 Daltile
100–101 1 Andrew Twort/Redcover.com; 2 © Richard Bryant/Arcaid/Corbis; 3, 5 Carina Works; 4 Johnny Bouchier/Redcover.com
102–103 1 © Nancy Collinge/Fotolia.com; 2 © Lucas/Fotolia.com; 3 © Michael Travers/ Fotolia.com; 4, 7 © Rosita Fraguela/Fotolia. com; 5 © Alexander Sayganov/Fotolia.com; 6 © Paul-andrè Belle-isle/Fotolia.com; 8 © Emilia Kun/Fotolia.com; 9 © Peter Van Den Wyngaert/Fotolia.com; 10 © Vaida/Fotolia.com; 11 © NiceFoto/Fotolia.com; 12 © Andreas Gradin/Fotolia.com
104–105 Carina Works
106–107 1, 3 Andreas von Einsiedel/Designer: Anna and Aib Barwick; 2 Andreas von Einsiedel/Designer: Robert Hering; 4 Ashley Mossison/Redcover.com; 5 Andreas von Einsiedel/Designer: Juliet Bragg
108–117 Daltile
118–119 1, 3, 5 Indigenous; 2 Johnny Bouchier/ Redcover.com; 4 Jake Fitzjones/Redcover.com/ Designer: Chalon Kitchens
120–121 1, 7 © Anna Sirotina/Fotolia.com; 5 © Dariusz Gudowicz/Fotolia.com; 2–4, 6, 8 Mandarin Stone; 9–16 Bettini Tile Service
128–129 Daltile
130 Kährs
131 Mullican Flooring
132–133 1, 4, 5 Quick-Step; 2, 3 Kährs; 6 Henry Wilson/Redcover.com
134–135 Kährs
136–137 1 Mannington; 2 Award Hardwood Floors; 3–12 Kährs
138–139 1–4 Teragren; 5–8, 10-12 Kährs; 9 Mannington
140–141 1, 4, 9 Anderson Floors;

2 Mannington; 3, 10–12 Award Floors; 5–8 Kährs
142–143 1–4, 6–12 Kährs; 5 Anderson Floors
144–145 1–2, 5, 8, 10 Kährs; 3, 7, 9 Anderson Flooring; 4 Mannington; 6, 11, 12 Award Floors
146–147 1 Duro Design; 2, 3, 5, 6 Kährs; 4 Award Floors
148–149 1, 3, 4 Lauzon, 2 Henry Wilson/ Redcover.com; 5 Reto Guntli/Redcover.com
150–151 1 Kährs, 2, 3 Quick-Step; 4–11 Jonathan Baker; 12 Real Hardwood Floors
152–153 1, 2, 5 Pergo; 3, 4 Quick-Step
154–155 Quick-Step
156–157 1 Huntley Hedworth/Redcover.com; 2 Peter Woloszyski/Redcover.com; 3 Andreas von Einsiedel/Architect: Hugh Newell Jacobsen, FAIA; 4 Johnny Bouchier/Redcover.com; 5–8 Teragren
158 Andreas von Einsiedel/Designer: Monique Waque
164–165 Kährs
166 Forbo
167 Armstrong
168–169 1 Duro Design; 2, 3, 5 Armstrong; 4, 6 Forbo
170–171 1, 2, 4, 5 Duro Design; 3 Amorim
172–173 Duro Design
174–175 1, 3–5 Interior Leather Surfaces; 2 Winfried Heinze/Redcover.com
176–177 Interior Leather Surfaces
178–179 1 Forbo; 2, 3, 5 Congoleum; 4 Armstrong
180–181 swatch fan Forbo; 1–12 Armstrong
182–183 Congoleum
184–185 1 Richard Powers/Redcover.com; 2 Ed Reeve/Redcover.com; 3 David Giles/EWA Stock
186–187 1–4 © Nicole Waring/Fotolia.com; 5, 6 © Roman Sigaev/Fotolia.com; 7, 8, 12 © Adam Borkowski/Fotolia.com; 9, 10 © matttilda/Fotolia.com; 11 © EuToch/ Fotolia.com
188–193 Armstrong
194–195 1–4 Forbo, 5–8 Armstrong
196 Merida Meridian
197 Huntley HedworthRedcover.com
198–199 1, 3 Nourison; 2, 6 Merida Meridian; 4 Mike Daines/Redcover.com/Designer: Jo Warman; 5 Villa Nova
200–201 1 Tim Evan Cook/Redcover.com; 2 Sophie Munro/Redcover.com; 3 © Jason Stitt/ Fotolia.com; 4 Jean Maurice/Redcover.com; 5 © Glenn Sandul/Fotolia.com; 6 Jake Fitzjones/Redcover.com; 7 Reto Guntli/ Redcover.com
202–209 Karastan
210–217 Merida Meridian
218–219 1, 3, 4 Nourison; 2, 6 Merida Meridian; 5 Christine Bauer/Redcover.com
220–221 1, 3–10 Nourison; 2 The Rug Company
222–223 1–4 Nourison; 5, 8 GuildCraft Carpets; 6, 7, 9, 10 The Rug Company
224–225 Merida Meridian
226–227 1–4 Warwick; 5–8 Merida Meridian